Water Quality Management: How to Structure it Within a Utility

The mission of the Awwa Research Foundation (AwwaRF) is to advance the science of water to improve the quality of life. Funded primarily through annual subscription payments from over 1,000 utilities, consulting firms, and manufacturers in North America and abroad, AwwaRF sponsors research on all aspects of drinking water, including supply and resources, treatment, monitoring and analysis, distribution, management, and health effects.

From its headquarters in Denver, Colorado, the AwwaRF staff directs and supports the efforts of over 800 volunteers, who are the heart of the research program. These volunteers, serving on various boards and committees, use their expertise to select and monitor research studies to benefit the entire drinking water community.

Research findings are disseminated through a number of technology transfer activities, including research reports, conferences, videotape summaries, and periodicals.

Water Quality Management: How to Structure it Within a Utility

Prepared by:
Edward G. Means III
McGuire Environmental Consultants, Inc.
3471 Via Lido, Suite 207, Newport Beach, CA 92663
and
Mark LeChevallier
American Water Works Services Company, Inc.
1025 Laurel Oak Road, P.O. Box 1770, Voorhees, NJ 08043-1770

Sponsored by:
Awwa Research Foundation
6666 West Quincy Avenue, Denver, CO 80235-3098

Published by:

Awwa Research Foundation
Advancing the Science of Water®

IWA Publishing

DISCLAIMER

This study was funded by the Awwa Research Foundation (AwwaRF). AwwaRF assumes no responsibility for the content of the research study reported in this publication or for the opinions or statements of fact expressed in the report. The mention of trade names for commercial products does not represent or imply the approval or endorsement of AwwaRF. This report is presented solely for informational purposes.

Published by IWA Publishing, Alliance House, 12 Caxton Street, London SW1H 0QS, UK

Telephone: +44 (0) 20 7654 5500; Fax: +44 (0) 20 7654 5555; Email: publications@iwap.co.uk;
Web: **www.iwapublishing.com**

AwwaRF report number 91007F. Originally published by AwwaRF for its subscribers in 2004.
IWA Publishing version published 2005.

ISBN 1 84339 902 4

CONTENTS

TABLES

LIST OF FIGURES

FOREWORD

The Awwa Research Foundation is a nonprofit corporation that is dedicated to the implementation of a research effort to help utilities respond to regulatory requirements and traditional high-priority concerns of the industry. The research agenda is developed through a process of consultation with subscribers and drinking water working professionals. Under the umbrella of a Strategic Research Plan, the Research Advisory Council prioritizes the suggested projects based upon current and future needs, applicability, and past work; the recommendations are forwarded to the Board of Trustees for final selection. The foundation also sponsors research projects through the unsolicited proposal process; the Collaborative Research, Research Applications, and Tailored Collaboration programs; and various joint research efforts with organizations such as the U.S. Environmental Protection Agency, the U.S. Bureau of Reclamation, and the Association of California Water Agencies.

This publication is a result of one of these sponsored studies, and it is hoped that its findings will be applied in communities throughout the world. The following report serves not only as a means of communicating the results of the water industry's centralized research program, but also as a tool to enlist the further support of the nonmember utilities and individuals.

Projects are managed closely from their inception to the final report by the foundation's staff and a large cadre of volunteers who willingly contribute their time and expertise. The foundation serves a planning and management function and awards contracts to other institutions such as water utilities, universities, and engineering firms. The funding for this research effort comes primarily from the Subscription Program, through which water utilities subscribe to the research program and make an annual payment proportionate to the volume of water they deliver and consultants and manufacturers subscribe based on their annual billings. The program offers a cost-effective and fair method for funding research in the public interest.

A broad spectrum of water supply issues is addressed by the foundation's research agenda: resources, treatment and operations, distribution and storage, water quality and analysis, toxicology, economics, and management. The ultimate purpose of the coordinated effort is to assist water suppliers to provide the highest possible quality of water economically and reliably. The true benefits are realized when the results are implemented at the utility level. The foundation's trustees are pleased to offer this publication as a contribution toward that end.

Walter J. Bishop
Chair, Board of Trustees
Awwa Research Foundation

James F. Manwaring, P.E.
Executive Director
Awwa Research Foundation

ACKNOWLEDGMENTS

East Bay Municipal Utility District deserves foremost recognition for their kind hosting of the Expert Workshop. The project could not have been accomplished without their generosity and the special hospitality of their staff (especially Ron Hunsinger, Andy Enos, Ana Reyes, and Ken Pent).

The authors of this report are indebted to the following outstanding water quality leaders for their support, hard work and volunteer participation in the Expert Workshop:

Michael Barsotti - Champlain Water District
Miranda Brown – Washington Suburban Sanitary Commission
Ed Cummings – Zone 7 Water Agency
Ian Douglas – City of Otawwa
Bill Gedney – Southern California Water Company
David Hilmoe – Seattle Public Utilities
Mark Knudson – City of Portland
Gary Lynch – Park Water Company
Douglas Neden – Greater Vancouver Regional District
Sandy Oblonsky – Santa Clara Valley Water District
Pankaj Parekh – Los Angeles Department of Water and Power
Dan Quintanar – Tucson Water Department
John Rudnickus – City of Toronto
Sylvia Barrett – Metropolitan Water District of Southern California
Andy Enos – East Bay Municipal Utility District

Our Project Advisory Committee, who provided important guidance, insight, and support throughout the project as well as participated in the Expert Workshop, is especially thanked:
Ron Hunsinger – East Bay Municipal Utilities District
Barry MacBride – City of Winnipeg
Christine Owen – Tampa Bay Water

Special thanks goes to the project team including Lorena Ospina who coordinated the workshop, helped prepare the workshop materials and provided significant administrative and technical input into the project, Liz Freedman who helped with administrative matters and report formatting, Mike McGuire and Michelle Frey, who helped facilitate the Expert Workshop and Ken Reich who assisted in preparation of the workshop materials and helped facilitate the Expert Workshop. Michelle provided valuable assistance in articulating the foundational support systems.

Finally, the AwwaRF Project Manager, Ms. Elizabeth Kawczynski, lent her water quality expertise and project management insight in guiding the project. Thank you Elizabeth.

EXECUTIVE SUMMARY

RESEARCH

The objective of this project was to identify the key elements of successful management of the water quality function in a water utility and identify steps water utilities can take to improve their water quality management performance.

APPROACH

Water quality management has many dimensions including operations support, laboratory management, project planning, data management, research, technological applications, staff development, customer satisfaction, budget management, technological advancements and regulatory compliance. Water utilities have developed pragmatic organizational and strategic methods to manage these responsibilities, and their water quality managers have acquired substantial related expertise. This experience can be used by other water utilities seeking to improve their water quality management performance.

The Project Team's approach involved collecting available information through four means:

1. Examinations of available literature/experience on managing the "quality" function in organizations,
2. Developing and conducting a web-based survey of water quality managers to identify water quality organizational issues, and management issues.
3. Developing case studies to define strengths and weaknesses of water quality organizational models and
4. Conducting an Expert Workshop.

Areas of concern were identified in the web-survey and case studies. The information collected was synthesized into an issue paper. The issue paper served to frame the water quality management issues, identify measures of success or key organizational success factors for water quality management, and provide examples of successful organizational and functional approaches used by utilities that could be adopted or customized to improve water quality management. The issue paper also served as a basis for in-depth discussions at the Expert Workshop.

The Expert Workshop was designed to gather the wisdom and expertise of the participants by conducting thorough breakout group discussions. The participants included 17 utility volunteers, 3 AwwaRF Project Advisory Committee (PAC) members, 6 Project Team members from McGuire Environmental Consultants, Inc. and the American Water Works Services Company, Inc. The breakout group sessions addressed the following questions:

1. What are the goals of water quality management?
2. What are examples of successful organization and functional approaches?
3. What are measures of success for water quality management?

4. What questions should a utility ask itself in a self-assessment?

CONCLUSIONS

The literature review and web survey identified several key water quality management issues facing utilities today. Frequently cited issues included ineffective internal communications, lack of tools for evaluating water quality performance, data management and training constraints, and non-responsive decision-making processes. Some utilities have had some success in organizing water quality functions, however, many still find it difficult to match the dynamics of the industry and are resource constrained.

The workshop resulted in the definition of twelve functions water quality management should perform. These functions include:

- Compliance with Regulations
- Setting and Measuring Water Quality Goals
- Engaging in Water Quality Strategic Planning
- Managing Source Water Quality
- Managing Quality Through Treatment
- Managing Distribution System Water Quality
- Managing and Analyzing Water Quality Information
- Advocating and Training
- Interfacing with Regulatory Authorities
- Identifying, Managing and Achieving Community Expectations
- Emergency/Security Response

A thirteenth function, protecting public health, generated strong support and strong disagreement from the participants. Many suggested that protecting public health was the primary overall goal, an overarching function of water quality management.

These functions comprise an operating model that utilities can use to frame their water quality management strategy. That model is presented within this paper.

Specific goals were generated for each of these functions. These goals range from operating standards/targets to consumer outreach. In addition, the participants generated specific examples of water quality management success and "signposts of dysfunctionality" that are presented as a non-scored self-assessment in the appendices.

RECOMMENDATIONS

The study identified three broad requirements for water utilities to achieve sustainable water quality management excellence. These are:

1. Perform the twelve core functions of water quality management identified above,

2. Align the interests of customers/stakeholders, the governing board, general management, Water Quality Managers and the water quality organization,

3. Provide the "foundational systems" to support water quality management. These foundational systems include business systems, water quality knowledge management systems, performance management systems and communications systems.

4. Provide metrics to assure success in these functions.

Achieving sustainable water quality management excellence is facilitated when the utility:

1. ***Engages your governing board (community or public utilities commission or corporate)*** - Consciously evaluate the water quality posture of the organization. Understand the community's perspective on water quality and incorporate that perspective into the utility's decisions. Emphasize the organization's desired water quality posture in the utility's strategic plan. Communicate that commitment throughout the organization.

2. ***Documents management support*** - Encourage senior management to establish clear goals/objectives/strategies for water quality management and communicate/advocate the importance of success throughout the organization.

3. ***Sets water quality goals and monitors those goals*** - Establish water quality goals and monitor progress towards achieving those goals. The goals should be understood and (ideally) developed by consensus of the organization.

4. ***Develops an advocate*** - Seek to develop a respected leader responsible for water quality. Recognize the dual, audit/assistance role of Water Quality in relationship to Operations. Work to develop and maintain trust between the water quality technical support function and the operations group. Communicate. Position the Water Quality function to be viewed as a resource to the organization.

5. ***Documents responsibilities*** - Define and document the water quality responsibilities of the organization. Ensure there is no confusion within the utility as to where that responsibility lies. Delegate the authority to act along with the responsibility.

6. ***Provides timely information*** - Routinely generate water quality information that is easily understood. Broadly disseminated that information internally and externally. Share information with your customers and stakeholders.

7. ***Stays current*** - The pace of technological change (process technology, analytical equipment, information systems, on-line monitoring capability, regulatory developments, etc.) is rapid. Stay current through training, conference attendance, publications, trade associations, etc. This requires the necessary organizational commitment to provide adequate budget and staffing.

CHAPTER 1
INTRODUCTION

BACKGROUND

"Water quality" is used throughout this document. Water Quality (capital letters) refers to the organizational unit in a utility that is responsible for managing the water quality function. Water quality (small case) is the *function.*

Water quality management has many dimensions including operations support, laboratory management, project planning, data management, research, technical applications, staff development, customer satisfaction, budget management, technological advancements and regulatory compliance. Water utilities have developed pragmatic organizational and strategic methods to manage these responsibilities, and their water quality managers have acquired substantial related experience. This experience can be used by other water utilities seeking to improve their water quality management performance. The goal of this project was to document key organization success factors for water quality management, provide examples of successful approaches and outline a process that can be used to evaluate needs and develop an optimal organizational strategy.

APPROACH

The project approach involved a collection of available information through four means:

1. Examination available literature/experience on managing the "quality" function in organizations.
2. Developing and conducting a web-based survey of water quality managers to identify water quality organizational issues,
3. Developing case studies exploring water quality organizational models, strengths and weaknesses, and
4. Conduct an expert workshop.

The information from the first three sources was synthesized into an issue paper and formed the basis of dialogue at the Expert Workshop. The purpose of the workshop was to solicit and document the collective water quality management wisdom of some of the top water quality and management staff in the country. Through facilitated sessions, the Project Team identified successful water quality management practices and measures of water quality management success. The supporting information for the workshop is included in the Appendices.

WATER QUALITY MANAGEMENT WORKSHOP

The purpose of the Expert Workshop was to involve a representative group of water quality managers, operations staff, general managers, engineering staff and others in a dialogue on effective water quality management function within utilities. The list of participants includes approximately 20 utility volunteers, 3 Project Advisory Committee members and 6 Project Team members. Participants are identified in Appendix D. The workshop was held on August 15-16, 2002 in Northern California.

Prior to the workshop, participants were asked to review the results of the web survey and case studies as they were presented in the issue paper prepared for the meeting. In addition, they were asked to spend some time contemplating the following four questions:

1) What are the goals of water quality management?
2) What are examples of successful organization and functional approaches?
3) What are the measures of success for water quality management?
4) What questions should a utility ask itself in a self-assessment?

Day 1 of the workshop began with an introduction and presentations from selected speakers. The purpose of the presentations was to frame current water quality management issues and practices. The presentations were followed by a session of breakout group discussions and report backs to review and define "water quality management goals."

The second day of the workshop began with a review of the previous day's input. Breakout group sessions continued throughout the day. The goal of the Day 2 morning breakout sessions was to identify examples of water quality management success. The afternoon sessions focused on developing measures of success and self-assessment questions utilities could use to measure their success.

CHAPTER 2
WATER QUALITY MANAGEMENT FUNCTIONS

Successful water quality management is associated with the fulfillment of twelve general "roles". These roles will vary by water utility; there is not one, best, way to manage this function. Water utilities can emulate these models to improve their water quality management effectiveness.

The Expert Workshop participants covered a wide range of functions that a progressive water quality management unit will carry out. These functions have been consolidated into 11 areas. These include:

- Compliance With Regulations
- Setting and Measuring Water Quality Goals
- Engaging in Water Quality Strategic Planning
- Managing Source Water Quality
- Managing Treatment
- Managing Distribution System Water Quality
- Managing and Analyzing Water Quality Information
- Advocating and Training
- Interfacing with Regulatory Authorities
- Identifying, Managing, and Achieving Community Expectations
- Emergency/Security Response
- Public Health
- Protecting

These functions are discussed individually below. The discussion includes specific examples of workshop-developed goals that water utilities can consider for each functional area.

COMPLIANCE WITH REGULATIONS

A fundamental function of water quality management within water utilities is to ensure that the utility consistently complies with drinking water regulations (federal, state and local). Compliance with these regulations requires a commitment to understanding them, devising the monitoring programs to ascertain compliance and consistently meeting the reporting requirements of the regulations. Several points are implicit in carrying out this function.

- The entire organization needs to share an understanding of the importance of compliance.
- There needs to be a clear delineation of authority for compliance. No confusion between the operations function and the water quality function should exist. The planning, finance and public relations functions must value water quality. Senior Management should establish this "tone". Inter-organizational competition for resources can quickly reduce the utility focus on water quality.

- The information management systems necessary for compliance must be in place. This requires access to modern data management systems and decision support systems. Rapid dissemination of accurate information is critical. Many participants in the workshop cited their relatively primitive water quality data management capability. This is remarkable as the invited participants generally represent some of the best performing utilities in the country.
- Clear organizational accountability for non-compliance must be articulated.

Example Goals include:

- There will be no regulatory violations.
- We will proactively optimize our water treatment processes.
- We will produce accurate and timely water quality information that supports the cost-effective management of water quality.
- We will detect developing problems in time to correct them.
- We will clearly define the chain of responsibility and accountability

SETTING AND MEASURING WATER QUALITY GOALS

In addition to consistent compliance with primary and secondary water quality standards, progressive water quality management/utilities establish and track conformance with internal water quality goals. These goals typically "push the envelope". They are designed to instill a sense that regulatory standards are the minimum acceptable performance and that "we can do better" than those requirements in terms of providing water quality. The goals might entail "action levels" to trigger operational responses to avoid pending non-compliance. They might include goals that reflect community desires to minimize brominated disinfection by-products. The goals may or may not consider cost issues (e.g. treatment chemical usage). Specific examples of goals in some of the Case Studies are included in Appendices.

It is clear from reviewing the Case Studies and talking to the workshop participants that the relationship between operations and water quality is critical and that there exists numerous management structures that share/split responsibility for meeting water quality goals and regulation. Establishing clear goals and providing technical direction and training to operators and other operational staff to be able to meet the goals is a desirable function of water quality management. It serves two roles: 1) it builds a cooperative relationship with operations that is essential to high performance water quality management and 2) it elevates the stature of the water quality function within the organization.

Example Goals include:

- We will develop and document water quality goals for the utility.
- The goals will be understood and shared by the utility employees.
- The goals will reflect customer expectations.
- We will establish monitoring programs to track performance towards the goals.
- The goals will be met in a cost effective fashion.
- Warning limits (action levels) will be established as indictors of pending non-compliance

ENGAGING IN WATER QUALITY STRATEGIC PLANNING

Progressive water quality management is proactive in its activities whether they be developing inter-organizational action plans for compliance with a regulation or establishing budget priorities. Progressive water quality management is included in utility decision-making. This requires that the utility recognize the importance of water quality. Providing leadership within the utility in the water quality area requires the water quality manager to think strategically about the role of water quality within the utility and the external drivers affecting the water quality posture of the utility. A function of water quality management is to lead and coordinate the utility's water quality strategic plan (or the water quality elements of the organization's strategic plan).

Example Goals include:

- We will develop a 5-year iterative water quality strategic plan (WQSP).
- The water quality strategic plan will support the attainment of water quality goals.
- Water quality management will support the corporate strategic planning efforts.
- Specific water quality language will be included in the utility's strategic plan.
- We will work to ensure that all project development and implementation functions identify impacts on water quality.

MANAGING SOURCE WATER QUALITY

The management of water quality historically focused on the water treatment plant. The passage of the 1996 amendments to the Safe Drinking Water Act began a turn toward managing water quality contaminant contributions in the rivers and streams that provide water supply to the nation's water utilities. The Clean Water Act and the Farm Bill have furthered the development of watershed management strategies to reduce pollutant loading on water treatment plants. Progressive water utilities are leveraging the tools in legislation/regulation to reduce the need for expensive water treatment plant upgrades.

Example Goals include:

- We will protect our source water quality.
- We will provide early warning of potential problems.
- We will proactively seek to influence land use policies in our watersheds.
- We will influence human behavior that contributes source water contamination.

TREATMENT

It is clear from reviewing the Case Studies and talking to the workshop participants that the relationship between operations and water quality is critical and that there exists numerous management structures that delineatet responsibility for meeting water quality goals and regulation. Establishing clear goals and providing technical direction and training to operators and other operational staff to be able to meet the goals is a desirable function of water quality management. It serves two roles: 1) it builds a cooperative relationship with operations that is essential to high performance water quality management and 2) it elevates the stature of the water quality function within the organization.

Example goals include:

- Include phased "trip wires" - Consider multiple escalating targets that trigger different actions. (e.g. particle counts as warning for turbidity >3 um, >10 um, <0.1 NTU all filters).
- Seek consistency of operations
- 100% of the time <0.20 NTU
- 95% of the time <0.15 NTU
- Try to keep the goals simple and understandable.
- Participation in the Partnership for Safe Water.

MANAGING DISTRIBUTION SYSTEM WATER QUALITY

Distribution system water quality historically has received less focused attention than it will in the future. USEPA is developing distribution system water quality regulations that will likely place additional demands on water utilities. From a practical standpoint, water treatment has largely been optimized and source water protection is underway, leaving the distribution system as the remaining opportunity for improving the quality of water delivered to consumers. Progressive water quality management will actively monitor and assess distribution water quality, and will likely do so using real-time monitoring systems.

Example Goals include:

- We will integrate water quality and hydraulic goals into the operation of the distribution system.
- We will develop, implement and maintain a distribution system water quality maintenance plan, including flushing, tank cleaning, etc.
- We will rapidly respond to and correct any distribution system water quality deterioration.
- We will provide comparable water quality to all customers.
- We will monitor and understand water quality in the distribution system.

MANAGING AND ANALYZING WATER QUALITY INFORMATION

Participants in the workshop repeatedly noted the large quantity of data that water quality managers must assimilate and interpret. Some noted that the water quality function has become so specialized that some laboratory analysts do not recognize the operational significance of the data they are generating. They generate volumes of data rather than interpret that data and generate information. Accordingly, progressive water quality management will put in information systems that can simplify the generation of "actionable" information. These systems will integrate laboratory information management systems, supervisory control and data acquisition systems field data and customer complaint systems with regulatory levels and water quality goals.

Example Goals include:

- To provide useful, reliable, accurate, contextual and timely water quality information.
- This information will be provided to all utility functional areas that require such actionable information in a manner that is meaningful to their function within the utility.

ADVOCATING AND TRAINING

Effective water quality management presumes the necessary basic skill sets exist within the utility's organization. Among these skills is the capability to generate, interpret, and manage water quality information. It also presumes leadership within the organization. Ideally, water quality managers are technically proficient, sought out by the organization as a resource, and engaged in the "business" of the utility. Providing water quality leadership within the utility is an important function of effective water quality management. Related to this attribute is the function of providing the required operational and regulatory training to enable sound utility decision-making on matters that affect, or are affected by, water quality issues. This builds the water quality "ethic" of the entire organization. Progressive water quality management requires that the utility remain current with regulations and trends. This is facilitated through conference attendance and subsequent reporting back results of the presentations to staff and to senior management to underscore the value of attendance. The development of young scientists and engineers is also enhanced through participation in such training events.

Example Goals include:

- We will be aware of compliance trends, including new regulations.
- We will maintain regular communications, while building trust and confidence.
- We will be involved as a productive and constructive resource in decision-making that affects, or is affected by, water quality.
- We will ensure alignment between water quality policy and implementation.
- We will acquire the necessary resources to meet the organization's water quality objectives.
- We will support utility management in meeting the overall mission.
- We will provide inter-organizational training to improve the utility's knowledge of water quality, drinking water regulations and treatment technology.

INTERFACING WITH REGULATORY AUTHORITIES

The effective management of water quality requires the development and maintenance of professional, respected and trusted relationships with the local regulatory authorities. Developing open channels of communication between the utility and the regulator helps avoid destructive confrontations. A close working relationship between the regulator and the utility helps avoid community perceptions that the utility is disengaged. Given this, it is vitally critical that specific lines of communication be established within the utility. The "regulatory liaison" should not be ambiguous. Communications with the regulator should be directed through this liaison. Confusion occurs when there are multiple contacts with the regulators. The contact protocol should be defined.

Example Goals include:

- We will positively represent a utility's view and contribute to better regulation.
- We will assist the regulators in reaching an optimal outcome.
- We will communicate with the regulators routinely.
- We will avoid "surprising" the regulator.
- We will build a professional relationship with the regulator.

- We will develop a climate of mutual respect and cooperation with regulator.
- We will seek early involvement of the regulator in important water quality decisions.

IDENTIFYING, MANAGING, AND ACHIEVING COMMUNITY EXPECTATIONS

A valuable role or function that Water Quality management can play in a utility is to be an internal voice for the community's views on water quality so these views are conveyed to management and the governing board. Frequently, governing boards must make difficult policy decisions involving balancing of costs and health effects. Understanding what the community believes, perceives and desires is vital to allowing a politically sustainable decision on a water quality issue. Such community dialogue may increase the complexity of the actual decision but it will allow the gauging of such community issues as values, willingness to pay, affordability, and other concerns. This makes for better decisions and better government. In addition, it allows management to anticipate areas of controversy. The decision may not ultimately be embraced by the community, but there should be no surprise as to the community reaction.

Example Goals include:

- We will acquire, interpret and share community water quality perspectives.
- We will provide accurate and timely customer perspectives to utility decision makers.
- We will apply the understanding of customer perspectives to the development and revision of water quality goals.
- We will seek constant improvement in customer satisfaction regarding water quality.
- We will reduce complaints, while increasing overall customer satisfaction.
- We will understand customer needs for water quality information and provide that information.

EMERGENCY/SECURITY RESPONSE

Water quality management has a critical role in the effective and timely response to emergencies. Specific strategies should be developed to provide the technical resources for water quality responses to contamination events.

Example Goals include:

- We will ensure the customer understands the public health/perception issues related to contamination events.
- We will consider and address appropriate vulnerabilities and clearly identify communication plans, responsibilities and accountabilities.

PROTECTING PUBLIC HEALTH

A function that received substantial discussion and for which there was no clear majority view involved the role of the utility in protecting the public health. Numerous participants feel that progressive water quality management structures internalize the role of "public health protector." Others believe that that role belongs to the regulator and the medical community with the utility playing a more passive role. Utilities that choose an active role can establish this goal as a guiding principle in the conduct of utility business; the goal can share equal importance with providing adequate supplies of water. The utility can go beyond the regulatory requirements to inform the public regarding the effects of water quality. It can actively gauge the concerns of its customers and craft programs to respond to those concerns.

Example Goals include:

- Our utility will proactively address water quality regulations, consumer preferences and perceptions.
- We will provide timely dispensation of water quality information to customers and stakeholders.
- We will be viewed as an advocate for water quality in the community.

These twelve functional responsibilities of effective water quality management can form the core elements of water quality strategic plan. Each utility can tailor their plan to local conditions/ resource constraints.

CHAPTER 3
THE WATER QUALITY MANAGEMENT CONSTRUCT

WATER QUALITY MANAGEMENT INTERACTIONS

Water quality management interactions can be characterized as:

- Extra-organizational – interactions outside of the utility (e.g. public, regulator, legislator, etc.),
- Inter-organizational – interactions with other utility units (e.g. operations, engineering, finance, public relations, etc.), and
- Intra-organizational. – interactions within the water quality function (e.g. laboratory operations, samplers, quality assurance, report preparation, etc.).

Extra-organizational Interactions

Extra-organizational interactions are a critical part of effective water quality management (and, arguably, an area where water utilities could improve). These interactions are less confrontational when the utility understands (and ideally, shares) the culture, norms, mores, values and expectations of the customer base. These beliefs are the result of the life experiences of the population served by the utility colored by current economic and political realities. These beliefs (which define issues of willingness to pay, affordability, environmental ethic, priorities, etc.) should be embodied in the *Mission* and *Goals* of the utility through the governing body. These bodies may be boards, commissions, councils, etc.

Questions that deal with extra-organizational issues include:

- Do the actions of the utility (water quality or otherwise) reflect the values of the community?
- Does the governing body reflect the values of the community?
- Is there open and regular communication between the utility and the community?
- Does the community receive adequate information as to the activities of the utility?
- Is the community engaged?
- Are the community's values reflected in the governing board's deliberations?
- Are the community/governing board values reflected in the utility policies?

Establishing the community's perspective on water quality is a first step to ensuring the utility's policies reflect the community's values. Once the utility policies are aligned with the community's values, management can carry out the policies with reasonable certainty that they have community support. Regular calibration of those values/policies ensures that the utility is never far from the community values. Survey, focus groups, complaint forums, etc. are ways to solicit this information.

Inter-organizational interactions

Inter-organizational interactions occur between the water quality management function with the other functional areas of a water utility. These other functional areas (they are called different things in different utilities) include:

- General management
- Operations
- Maintenance
- Engineering
- Planning and water resources
- Finance
- Information technology
- Public and government affairs
- Regulatory affairs
- Audit
- Legal
- Right of way
- Customer service

Each of these functional areas has a role and responsibility/accountability that may or may not be aligned with the water quality management function (or vice-versa). Aligning these respective interests is critical to effective water quality management (as well as general utility success).

Clear definition of responsibility and authority is fundamental to effective functioning in organizations. This is also the case for water quality management. The absence of definition degrades the quality of the decision making process and the decisions.

Questions that deal with *inter-organizational* interactions include:

- What policies or goals govern water quality management?
- Are they documented?
- Are they understood and shared by the Board, commission, management and rank and file?
- How do the organizational units communicate?
- How effective is that communication?
- Does the communication function in emergencies?
- Where does the water quality decision-making authority lie?
- Is water quality an audit or an assistance function? Or both?
- Is the utility positioned to influence and respond to regulations/legislation? Is it important to be so positioned?

Intra-organizational Functions

The *Intra-organizational* Water Quality Management relates to the day to day "business" of water quality and involves such questions as:

- Is there strong leadership within Water Quality:
- Is sample collection, analysis and data management/reporting purposeful/efficient?
- Are water quality resources deployed appropriately?
- Are the quality control and quality assurance functions robust?
- Does the water quality function have the required resources to carry out its responsibilities?
- Does the water quality unit work as a team?
- Does staff have the necessary training/resources to fulfill their responsibilities?
- If the utility conducts research, are there formal research plans, objectives, budgets and schedules?
- Are the research initiatives focused to answer the right questions and actively managed?
- Do staff have a clear understanding of their roles, responsibilities and accountabilities?

ALIGNING WATER QUALITY STAKEHOLDERS

There appears to be an opportunity to positively effect water quality management performance by increasing stakeholder involvement. Over half of the survey respondents believed that the water quality function was appropriately integrated and influential in their organization. When asked about end customer involvement in the water quality decision-making process within the organization, about 40% indicated that the customer was rarely involved, and 50% indicated the customer was somewhat involved. Only 7% indicated that the customer was highly involved (although half thought that they should be). Other areas of concern included the lack of governing board interactions. Fifty percent of the participants indicated the governing board was not actively engaged in water quality issues, and 85% believed that they should be more engaged in these issues. These views were echoed in the Workshop.

In those water utility organizations which emphasize water quality, that emphasis typically comes top down; either the governing board or the general manager establishes the priority. Fostering an environment where water quality is highly valued requires strong leadership and a customer-centric attitude

Water Quality Alignment refers to the shared philosophy between the external stakeholders (community), the governing board, general management, water quality management and the water quality organization. Aligning these "levels" contributes to good water quality management performance.

The extra-, inter- and intra-organizational interactions described above occur can be aligned across five levels (Figure 3.1):

- External stakeholders
- Governing board
- General management

- Water quality management
- Water Quality staff

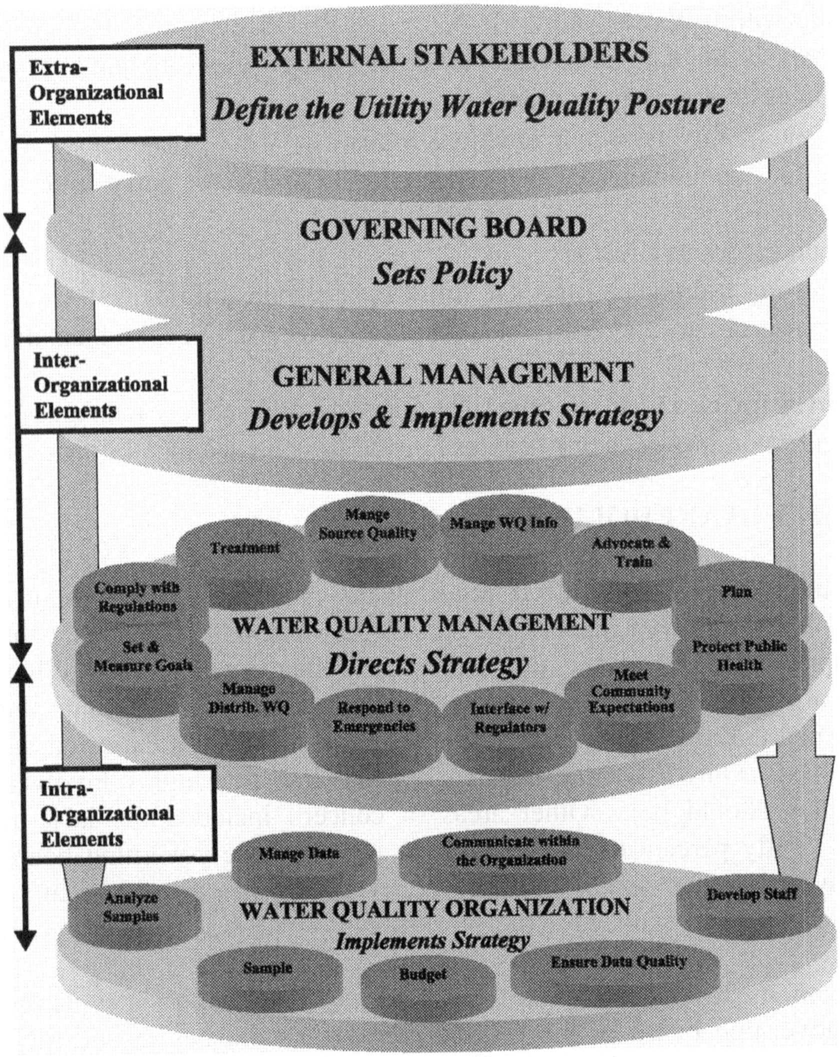

3.1 Model Water Quality Management Construct

The sustainable structuring of the water quality function requires certain things of each of the stakeholder groups:

Community Stakeholders

Community stakeholders include the consumer, the media, the business community, the environmental community, and others not associated with the utility. The public should be engaged in the activities of the water utility. In order for that to happen, the utility must take the initiative to reach out and make sure the public is informed regarding water quality issues. The collective "public" has common attitudes or philosophies about risk, cost, benefit, environmental posture, information needs, etc. Understanding these and how they may change over time allows the governing board to reflect the public's views in their actions. The "public" that shows up at the governing board meetings may, or may not, reflect the true attitude of the community. The utility must be proactive in gathering and understanding these attitudes and providing this information to the governing board.

Governing Board

Positive water quality management outcomes require support at the governing board. That support should (and, arguably, must) have the backing of the public (hence, the emphasis by many water quality leaders on consistently gauging public attitudes towards water quality and reflecting those attitudes in the utility's water quality policies and posture). The tone must be set at the top or the best of intentions from within the organization will not be sufficient to realize consistent excellence.

The governing board should establish the utility water quality philosophy through documented policies. These policies should reflect the public view. This philosophy should be incorporated in a formal strategic plan for the utility (mission statement, guiding principles, etc.). The strategic plan should establish water quality objectives for General Management to meet. Performance against these goals should be measured and executive compensation linked to that performance.

General Management

General Management must assist the governing board in understanding the regulatory requirements, costs, feasibility, trade-offs and, what the public views/perceptions are regarding water quality issues. General Management should work to:

- Fill information voids so that the adopted board policies are implementable,
- Assist the governing board by crafting and directing strategies to achieve the policy objectives,
- Establish and document the roles and responsibilities of the water quality management function in the utility,
- Hold managers accountable for failure to support the vision,
- Define the nature, frequency and content of decisional information required by senior management, and

- Provide the knowledge, business, communications and performance measurement systems to support the management of water quality (including training resources to ensure staff remains current).

Water Quality Management

Effective water quality management develops, analyzes and provides timely information to the organization (including the governing board). They are viewed as leaders in the organization and provide the "water quality conscience" to the utility management team. This requires maintaining a strong working relationship with the local regulatory agencies. Effective Water Quality management is proactive in its actions, following the "no surprises" rule. Water Quality management works cooperatively and proactively with other organizational units to analyze and develop response strategies for regulatory actions.

Effective management involves the integration of information from various departments and staff. Top-down or militaristic management can limit leveraging the knowledge assets of the organization. Collaborative management decision-making helps insure that vital strategic information is not overlooked through the exclusion of the affected functional areas from the decision-making process. However, top-down decision making definitely has its place during emergency situations.

Water Quality Organization

The water quality organization that executes utility water quality policy/strategy must be technically competent, current, proactive and responsive. The organization should be viewed as a resource/asset to the other organizational units in the utility. Timely and regular sharing of water quality data and information throughout the organization is important. Building trust within the organization requires communication and support.

The alignment of the community values, board policy, general management, Water Quality management, and water quality functional initiatives are essential to an effective and sustainable water quality management function in a utility. Failure to align these layers risks poor water quality performance (either perceived or real) and low stakeholder trust. Classic strategic planning can align these interests/perspectives.

Important questions that get at the heart of water quality alignment include:

- Does our governing board represent the community's views on water quality?
- What outreach do we conduct to ensure that?
- Is water quality management defined in terms of shared values, defined roles and responsibilities?
- Is water quality management involved in utility decisions?
- Does water quality management have access to the "decision makers"?

Capturing Public Confidence

Developing a sustainable and responsive water quality management function can raise public confidence in your utility. Figure 3.2 below depicts a model of public confidence interacting with water quality management.

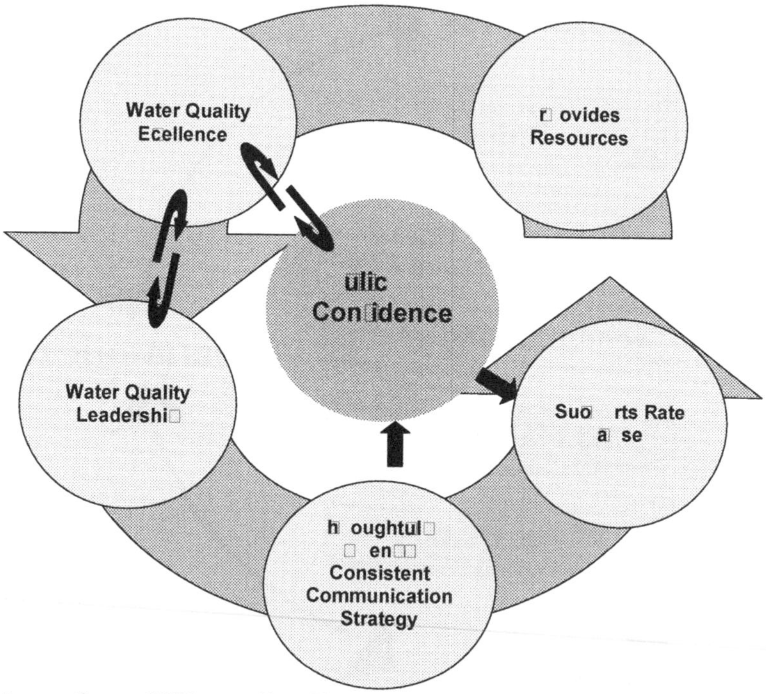

Figure 3.2 Interaction of Water Quality Excellence with Public Confidence

Water quality management success requires adequate resources to fund both the physical features (labs, water treatment plants, etc.) and the operational needs (staff, training, conference attendance, research, etc.). Adequate resources must come from an adequate rate base. An adequate rate base requires public support and confidence. Public support can be built through clear and consistent leadership that *believes* in water quality. Clear and consistent leadership requires a thoughtful and consistent communications strategy that is open and non-defensive. *This strategy must engage the public to be sustainable.* Breakdowns in any of these areas can diminish the effectiveness and responsiveness of the water quality management function.

FOUNDATIONAL SYSTEMS

The development of a successful water quality management function requires addressing the *twelve functions* and ensuring *water quality alignment of your utility.* There is a third element: establishing and maintaining what we term *foundational systems* (Figure 3.3). Foundational systems support water quality management. They distribute resources, knowledge, information and performance feedback. There are four basic areas of foundational systems:

- Utility business systems,
- Knowledge management systems

* Performance measurement systems; and
* Communication systems.

These systems support the five layers of stakeholders that comprise the water quality management construct (Figure 3.3 below).

Figure 3.3 Water Quality Management Foundational Systems

Utility Business Systems

Utility business systems include the financial systems that support the water quality staff as employees of the utility (payroll, budgeting, planning, legal, governing board, public relations, project management systems, financing etc).

Knowledge Management

Knowledge management systems capture and provide access to water quality information. These systems house the data and information assets of the water quality function and include supervisory control and data management systems, laboratory information management systems, security systems, operator logs, field logs, external data sources, customer complaint logs, etc. Integrating these systems remains a major challenge for water utilities.

The water quality management function in today's regulatory environment must assimilate and digest a huge volume of data. Few water quality managers appear completely satisfied with the nature, timeliness, accuracy and simplicity of the water quality decisional material they receive and transmit. Tracking tools are often cumbersome. Significant staff effort is required to compile and analyze current and historical data. Often historical data is still in paper records and little of the operational data from on-line monitoring may even be available making the utilities "data rich but information poor".

Many utilities are currently working to correct these problems, and are looking towards making investments in sophisticated integrated data management systems. Workflow, information retrieval and communication can be enhanced within an organization implementing novel and advanced applications that expand the limits of a traditional database (e.g. coupling graphical information systems with laboratory information management systems, supervisory control and data acquisition systems and the internet).

Defining and providing for the specific information needs of the stakeholders improves effectiveness and efficiency. These stakeholders include regulators (through regulatory reports), operations staff, consumers, management, media, and the water quality managers themselves. This knowledge allows us to define:

- what information is needed (e.g. what samples need to be collected),
- what form the information should appear in, and
- at what frequency the information is needed.

Ideally, stakeholders would get the *information* they need to make a decision (the system should allow them to get more if they want it) while avoiding information overload. *Implicit in achieving this goal is a fundamental review of the reasoning behind all "routine" reporting in terms of content, frequency and audience.*

Performance Measurement Systems

Performance monitoring systems routinely monitor and report performance against articulated standards and goals. These systems are for both "process" and "people".

Nearly all the participants surveyed responded that it was very important to establish clear water quality goals that are supported by senior management. These goals should be documented. They should be both numeric and descriptive. By implementing goals, an organization establishes a basis for measuring progress and aligning the interests of the organization. Involving all functional areas in setting the goals helps ensure organizational buy-in. The development of common goals by the utility stakeholders (operations, engineering, water quality, planning, etc.) can be done through typical strategic planning processes. Performance against these goals should be measured, reported and discussed.

A clear understanding of the specific roles and responsibilities of each of the functional and management areas within a water utility is also an important part of effective management of water quality. The water quality responsibilities of the organization should be institutionalized through a strategic planning process, operating policy and job descriptions. The responsibilities must be clearly communicated to the stakeholders (especially the inter-organizational stakeholders). There should be no doubt as to who is responsible for water quality regulatory compliance. Clearly, accountable staff must be provided the resources to succeed in carrying out the responsibility (regular training, equipment, etc.). Each layer of the management construct has a responsibility to the water quality management function. Accountability requires:

- a clear and shared understanding of the utility's water quality expectations,
- measurable goals and performance standards, and
- a system to track water quality monitoring and reporting information and provide "transparency" to management regarding performance.

Employee evaluations that gauge water quality performance should be performed on a regular basis. Some workshop participants cited the lack of a satisfactory evaluation system to measure performance and determine whether staff possesses the skill sets necessary to generate quality service and products. Participants also expressed concerns regarding the "lack of applicability" of standard evaluation forms to water quality functions.

There is a continuing need for a highly skilled workforce to meet increasingly complex utility operating requirements. Many survey respondents indicated that they did not have an adequate training program due to diminished budgets and resources. Approximately half of the utilities aggressively promote professional development (engaging water quality staff in 5 or more conferences/training events a year). In addition, approximately three quarters of the participants spend anywhere from 5 to 10 days per year in courses targeted to augment their professional development. For an organization to grow, investments must be made in the area of personnel development, both in professional, technical, and managerial training.

Communication Systems

Participants cited the need for effective communication. This was frequently brought up in the context of concerns over "organizational isolation". The isolation appeared to result from poor communications system fostering a sense of segregation. This can result in unclear roles and responsibilities for staff across organizational units, which can result in water quality management problems. Frequently, decisions regarding changes in water treatment were made without consulting the water quality group, even though they were directly impacted by the decision. Without an effective internal communications network there was no mechanism in place for water quality to voice their concerns.

Effective communication with the public was also found to be important to many utilities. Several identified programs to educate and listen to the public as an essential component of their success. Many employed the internet, publications, school fairs and surveys to gather customer concerns and perceptions regarding water quality. The lack of effective external communication strategies can hinder addressing public concerns and jeopardize productive "alignment" of values discussed earlier.

Effective water quality management is involved in developing strategic communications plans that proactively create and purposefully disseminate utility information/messages internally and externally.

STREAMLINING THE MANAGEMENT OF WATER QUALITY

There are multiple ways to tackle the task of streamlining. One extreme is to use a "zero based" budgeting approach and define the resources need for each of the twelve functional areas of water quality management described earlier.

A logical approach to a Water Quality Management review could follow the process depicted in Figure 3.4 below. The critical path is the establishment of specific functional goals.

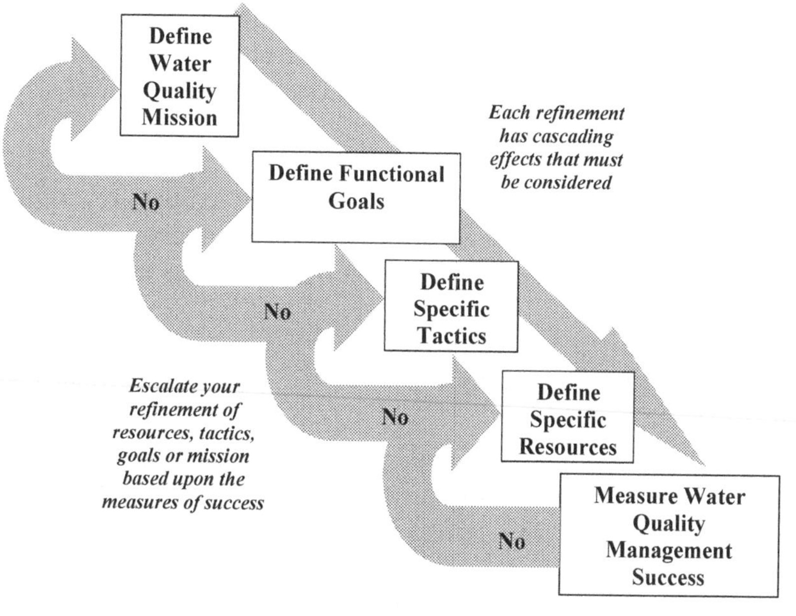

Figure 3.4 A Water Quality Management Review Process

Fundamental to a review of the water quality management function, is an understanding of what the organization "expects" from water quality. Generally this is established at a high level in the organizations strategic plan. The organization then establishes specific goals (and the water quality management function may have even more specific goals). The goals must have strategies and tactics defined. Then resources must be allocated and progress towards the goals measured.

This sequence of steps is iterative and time consuming. Like all utility processes, it works best when there is buy-in from the various stakeholders in the organization. This generally requires time and dialogue as water quality and public health risk issues are not viewed the same by all people. Experience, politics, wealth effects, age, health, philosophical issues, and other socio-political factors influence the attitudes.

The step that is likely under the greatest control of a water utility manager or water quality manager is the step that establishes "goals" for the utility. The twelve management functions described in Chapter 2 require specific goals; set them. Examples of goals are provided in the Appendices but each utility should create their own goals. Theses goals will define the posture of the utility and can serve as a benchmark against which resource needs can be assessed.

A possible process for streamlining the water quality management function then is to examine the critical systems/processes that must exist to achieve the adopted goals of the utility. Put another way, what specific tasks must be conducted or processes guaranteed to ensure the goal is met? Determining this requires a systematic, detailed examination of the current process and possible improvement of the process much as is done in other utility functional areas.

The resource allocation should be crafted to provide the minimum level of expenditure necessary to conduct the tasks to achieve the goal (progress against which is being measured). Failure to achieve the goal results in refining or even abandoning the resource expenditure, the tactic, the strategy, the goals or, in rare cases, the mission of the organization. These steps can be taken incrementally.

Where financial hurdles are great (which is in most utilities), a realistic revisiting of the tactics, strategies, and goals are required. The cost of some "industry leadership" activities may not be practical for the utility to assume. Or it may mean that the utility has to choose which areas it is going to "lead" in.

RESEARCH/SPECIAL INVESTIGATIONS

Conducting research in water utilities today can provide cost savings. Many staff members enjoy the variety and challenge of conducting process or applied research. Research programs can return far more than they cost (in savings due to process enhancements, avoided non-compliance, positive public recognition, regulatory influence, etc.) but they can also be expensive to run. Decisions to be a "leader" in research (or in a particular area) should be critically examined within the utility. The specific goal of "leadership" should be defined and measured. What is being sought? Is this for the utility's benefit or personnel/personal development? Or both? Defining and tracking success in this area will help boards and senior management continue to recognize the value of a research program to the utility. Participation in external research efforts (that don't require new internal laboratory capability) can be arranged through a number of organizations, including AwwaRF. One of the greatest benefits from participating in AwwaRF research is having outside experts examine your water quality issues. Through an active program, a manager can have worldwide experts providing their input to help address water quality improvements at significant cost savings. The American Water System has developed a solid rationale for the return on research/special investigations.

Water quality management is going to continue to be more important in the future - especially as regulations become more complex and intertwined. There are other water quality drivers that will elevate the importance of water quality management in the future including:

- Security/terrorism
- Greater customer expectations
- Infrastructure needs
- Treatment technology
- Laboratory capabilities
- Bottled water/POU treatment devices
- Source protection
- Global issues
- Emerging contaminants/detection enhancements
- Taste and odor
- Water reclamation

Water quality management success in the future can be achieved by critically evaluating the twelve functions describe in the report, aligning the stakeholders and providing the foundational system. This is a tall, but achievable prospect for water utilities.

APPENDIX A
BACKGROUND

IMPROVEMENTS IN WATER QUALITY

The water utility industry has made remarkable progress in meeting the requirements of the Safe Drinking Water Act. According to the Center for Disease Control the number of waterborne disease outbreaks in the U.S. appears to be declining. The number of outbreaks reported during 1997 (i.e., 7) and 1998 (i.e., 10) is comparable with those reported during 1996 (i.e., 6). However, the number of outbreaks reported for these 3 years is lower than those reported for any 2-year period since 1971. The increase and subsequent decrease in the number of reports might reflect, at least in part, changes in surveillance activities or be a reporting artifact. The recent decrease in the number of outbreaks reported since 1996 could also be caused by improved implementation of water treatment regulations (e.g., SWTR), increased efforts by many water utilities to produce drinking water substantially better than EPA standards requirements, and efforts by public health officials to improve drinking water quality.

These are rather impressive achievements in a comparatively short period of time. The drinking water community can take great pride in these gains, but these advances do not appear to be reflected in consumer perceptions.

CONSUMER BEHAVIOR

There have been numerous surveys conducted in recent years that shed light on the extent to which consumers have abandoned tap water and why. Most recently, the Water Quality Association published the 2001 National Consumer Water Quality Survey (February 2001). The national survey of 1021 adults found:

- Americans' awareness of and concerns about their drinking water quality is high. Eighty-six percent of respondents report some concern with their drinking water quality.
- Forty-nine percent of respondents believe federal drinking water laws are not strict enough (up from 40% in 1999).
- Home water treatment systems were used by 41% in 2001, compared to 38% in 1999 and 27% in 1995.

A June 2002 survey of 2029 California residents conducted by the Public Policy Institute of California (with a $\pm$2% margin of error) revealed:

- 24% drink tap water directly
- 35% filter tap water
- 39% drink bottled water
- 2 % answered other

WATER QUALITY MISSION STATEMENTS

The mission statements of most water utilities reveal core themes exemplified in the typical statement below.

"Our mission is to provide our service area with adequate and reliable supplies of high-quality water to meet present and future needs in an environmentally and economically responsible way."

or

"Our mission is to contribute to the well being of our community through the provision of an adequate supply of high quality drinking water"

The themes are typically focused on affordable, adequate and high quality supply. There are, of course, other strategic themes including environmental stewardship, open communication, customer focus, diversity, etc. but the core concepts of the mission statements are usually cost, quantity, and quality. The reality of water utility management is that there are trade-offs to be made in accomplishing the mission. Quantity and quality come with a price tag. Realizing those goals may not pose a "reasonable" cost for the utility. These cost trade-offs present some of the most difficult challenges to water quality management and to water quality managers. Viewed from a distance, there often appears to be a disconnect between what the public says they want and what water utilities often think they want and will pay for.

QUALITY MANAGEMENT BACKGROUND INFORMATION

There are several instructive efforts that can be used to guide the management of water quality in water utilities. These include:

1. ISO 9001:2000 standards,
2. The AwwaRF *Managerial Assessment of Water Quality and System Reliability* project, and the
3. *Australian Framework for the Management of Drinking Water Quality (2000)*

ISO 9001:2000

ISO is the International Organization for Standardization. ISO 9001:2000 specifies requirements for a quality management system for any organization that needs to demonstrate its ability to consistently provide product that meets customer and applicable regulatory requirements and aims to enhance customer satisfaction.

There are eight basic quality management principles on which the quality management system standards of the ISO 9001:2000 series (the latest version of the standard) are founded. These principles can be used by water utility management to improved performance. Indeed, top management involvement is required for the successful implementation of an ISO 9001:2000 system. The eight quality management principles include customer focus, leadership, involvement of people, process approach, systems approach to management, continual improvement, factual approach to decision-making and mutually beneficial supplier relationships.

Under the process approach, a quality management system is regarded as a single large process that considers many inputs to generate outputs. This large process has numerous internal, smaller processes that interact with each other. These include:

- Market Research Process
- Product Design Process
- Purchasing Process
- Production Process
- Service Provision Process
- Product Protection Process
- Customer Needs Assessment Process
- Customer Communications Process
- Internal Communications Process
- Document Control Process
- Record Keeping Process
- Planning Process
- Training Process
- Internal Audit Process
- Management Review Process
- Monitoring and Measuring Process
- Nonconformance Management Process
- Continual Improvement Process

In order to develop a quality management system that meets the new ISO 9001:2000 standard, it is necessary to:

1. Develop each process.
2. Document each process.
3. Implement each process.
4. Monitor each process.
5. Improve each process.

A detailed description of the ISO 9001:2000 Standard can be found on the ISO web site,

www.iso.ch/iso/en/iso9000-14000/iso9000/selection_use/selection_use.html.

MANAGERIAL ASSESSMENT OF WATER QUALITY AND SYSTEM RELIABILITY

More specific to the water industry than the general ISO Standards, AWWA studied water quality management as a portion of an investigation in the mid-1990s.

"Forty water utility leaders from across the United States gathered in an American Assembly, May 17-19, 1996, to record their perspectives on managing for water quality and system

reliability. They defined the challenge and identified strategic initiatives for utilities to apply over the next 10 years"

-American Assembly for AwwaRF "*Managerial Assessment of Water Quality and System Reliability*" project. (Cromwell et al. 1997)

The *Managerial Assessment of Water Quality and System Reliability* project examined the nuances of the provision of reliable and high quality water supplies from a management perspective. The project identified seven major driving forces that characterize the challenges in achieving water quality and system reliability:

1. **Customer relationships**

   The water industry must understand and be responsive to the needs of its customers."

2. **Funding issues**

   "There are numerous demands for limited funds which have encouraged water utilities to reduce cost through use of modern technology and optimum workforce strategies. One outcome of increased competition for scarce financial resources is that many utilities are now approaching their operations as a "business," seeking every cost-cutting and revenue-enhancing opportunity."

3. **Changing regulations**

   "The number and complexity of regulations have dramatically increased, along with the potential for regulatory violation. Changing regulations and scarcity of funds have put many managers in a speculative mode, waiting for regulatory certainty. The water industry should play a much more active role in monitoring and actively participating in the development of new regulations."

4. **Workforce issues**

   "The participants strongly believed that communication with, and the involvement of, employees in the decision-making process is vital."

5. **Changing technology**

   "The water industry manager must communicate these uncertainties to customers . . . and... proactively provide water quality data and interpretation."

6. **Availability of high-quality source water"**

   "The best quality sources have already been developed."

7. **Management/policy issues**

   "They (water quality managers) *must balance resource limitations with the need to meet increasingly stringent environmental regulations. They must develop and motivate employees while reducing operating costs. The dynamic nature of the issues they face requires continuous communication with both customers and policy makers. Given the complexity and number of issues, water utility managers must carefully build consensus on a set of priorities to get maximum return from the available financial resources. Water*

utility managers can leverage these limited resources through developing regional partnerships and non-traditional alliances."

The *Managing for Water Quality and System Reliability* project identified nine strategic initiatives to implement water quality and system reliability:

1. Involve the customer interactively to foster trust.
2. Develop a strategic plan.
3. Conduct strategic financial planning.
4. Develop a comprehensive quality management program to assure water quality and system reliability by implementing a customer-driven quality approach that aligns evaluation and compensation with reliability goals, and facilitates employee professional development and ownership. Drive decision-making down to the appropriate level in the organization.
5. Use advanced automation and information management technologies as a means of re-engineering work processes and rethinking communications and responsibilities.
6. Interact with legislative and regulatory groups.
7. Develop a comprehensive water quality management program that integrates source water, treatment, and distribution system reliability.
8. Develop and implement research programs to evaluate new solutions to existing and potential future water quality and reliability challenges.
9. Develop regional partnerships for cooperation and consensus building to obtain better reliability in source development, acquisition, and protection.

AUSTRALIAN FRAMEWORK FOR THE MANAGEMENT OF DRINKING WATER QUALITY

In the Australian Framework for the Management of Drinking Water Quality (*Framework*, 2001), established quality management practices were reviewed against the requirements for drinking water quality management.

The 12 primary points of the Framework are:

1. Commitment to Drinking Water Quality Management
2. Assessment of the Drinking Water Supply System
3. Planning-Preventive Strategies for Drinking Water Quality Management
4. Implementation-Operational Procedures and Process Control
5. Verification of Drinking Water Quality
6. Incident and Emergency Response
7. Employee Awareness and Training
8. Community Involvement and Awareness
9. Research and Development
10. Documentation and Reporting
11. Evaluation and Audit
12. Review and Continual Improvement

The Australian *Framework* was designed to apply to all water systems. It is necessarily generic and designed to be flexible. Current management of most drinking water supply systems will already incorporate many of the elements of the Framework but there is generally inadequate documentation/auditing for many of these preventive measures to ensure they are working properly. The *Framework* was designed to provide a higher level of assurance to regulators and the consuming public that comprehensive and responsive systems are in place and functioning to achieve consistently safe, good quality drinking water.

ADDITIONAL BACKGROUND MATERIAL

In addition to the three primary sources cited above, additional literature on the subject of quality management was reviewed and included in the issue paper for use in the Expert Forum. The information is briefly described below.

To manage effectively, you must communicate. But before you can communicate, you must speak a common language. Effective communication is the cornerstone of effective management, and learning how to establish effective communication channels throughout an organization is an important leadership skill (Aspan, 2000). This is especially difficult in water utilities as the high specialization of water quality professionals can increase the communication gap. Formal communication and decision-making strategies can facilitate water quality management.

Other business strategies for improving water quality management include involvement in benchmarking studies as a tool for organizational learning. Benchmarking studies have been identified as a method of performance improvement in the "learning organization." (Auluck, 2002). Excellence models and the self-assessment approaches can be employed to identify certain weaknesses.

A learning organization is one that is continuously transforming itself and is skilled at creating, acquiring and transferring knowledge to modify its behavior. (Garvin, 1993) Several authors highlight the following features as integral in a "learning organization":

- **Learning Climate** – the way the idea of learning is viewed within the organization; whether it is seen as functional and passive or developmental and integrated within the workplace. (Smith and Taylor, 2000)
- **Leadership** - leaders plays a key role in the development of the organization and they ultimately determine which behaviors are acceptable within the organization by setting an example. (Schein, 1997)
- **Communication** - By working as a team, learning from mistakes and continuous learning helps in the optimization of business functions. (IRS, 1998)

A notable amount of literature suggests that another component to business success is listening to customer demands. The component of a customers' perception of value cannot be undermined. Based on day-to-day interactions customers will use these experiences in evaluating the value of a product. Figure A.1 illustrates the components of customer value.

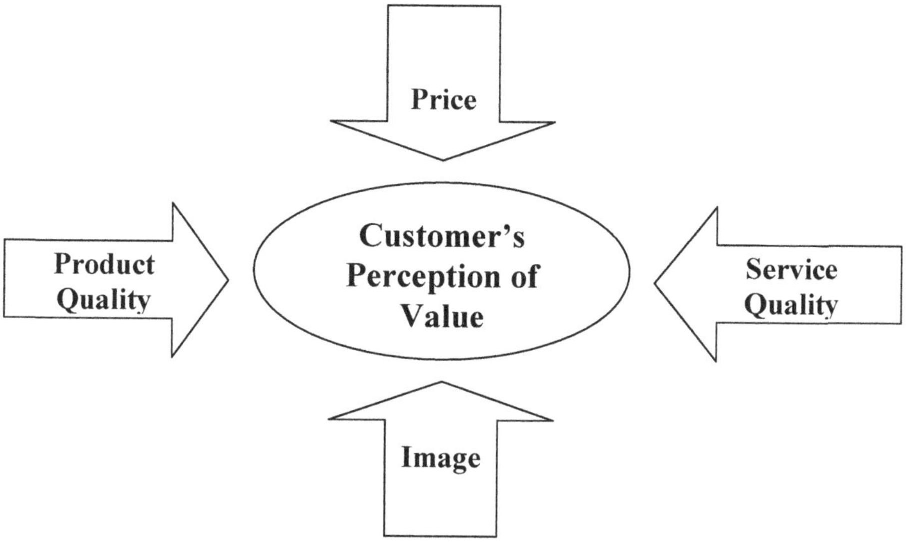

Figure A.1 Components of Value

Once a customer makes a decision regarding value, a fifth component of value emerges. That component is the relationship between the provider and the customer. Over time this relationship component can develop into an extremely important element. (Naumann, 2002). If the customer feels product or service quality is poor or the price is unreasonable the impression of "value" is low. What is uncertain is whether a low impression of drinking water quality "value" becomes reflected in the communities impression of the utility management, water rates, capital construction programs, etc. These are important questions as they speak to the potential for future "success" in the utility's endeavors.

Another essential tool of strategic management, especially in the operation of small to medium sized organizations, is that of networking. Managers often use their networking skills to identify and create value to their organization and to preserve and enhance competitiveness. (Gilmore, 2000).

A decision-support system should also be developed as a part of a highly integrated management system. This system provides organizations with the facilities for estimating future concerns within a business/utility. Based on an effective data management system, a manager can utilize this information to predict various water and resource demands. (Froukh, 2001). It is important for a business to maintain a highly effective and integrated data management system. The water utility community is generally data rich and information poor. This theme emerges in our case studies described later.

APPENDIX B
CASE STUDIES

Five utilities were interviewed as case studies. The case studies were designed to report and focus on organizational, management, personnel, communication, decision-making, and planning issues. The results were used to provide guidance for workshop deliberations. Interviews were carried out with water quality staff of the case study utility. Various utility sizes and geographic locations were selected to optimize representation. The case study interviews began the last week of May 2002.

The characteristics of the participating utilities are shown in Table B.1. The following five utilities participated in the case study evaluations:

Table B:1: Characteristics of Organizations in the Study

Utility	Service Location	Population Served
City of Ann Arbor	Ann Arbor, MI	10,000 – 50,000
American Water Works Service Co.	Nationwide	> 100,000
East Bay Municipal Water District	Oakland, CA	> 100,000
City of Ottawa	Ontario, Canada	> 100,000
City of Portland	Portland, Oregon	> 100,000

Each utility was asked to provide a water quality organizational chart prior to the interview. By collaboratively reviewing the chart structure with the participant, it facilitated the identification of gaps or barriers within the organization. In addition, the interviewee was asked to systematically review all roles and responsibilities affiliated with water quality. They were also asked to discuss management and decision-making practices.

CASE STUDY INTERVIEW GUIDE

Organization

- Please provide an organizational chart detailing the various job positions in your utility's water quality department (or in the department(s) where the water quality function resides). Please briefly describe all responsibilities associated with each job.

- How would you define the role of water quality in your utility? How would you define your role in the organization?

- From your experience which role(s) in your water quality department could be eliminated without affecting product quality.

- Which roles outlined in your organizational chart need major improvements?

- Which water quality role presents an economic burden/hardship in your utility?

- Does your utility have a water quality laboratory on the premises? Do you outsource any analytical work? If contract labs are used, how is data quality ascertained?

- Do you currently utilize a database system? If so, what is it mainly utilized for?

- How many regulatory violations have you had in the last 6 months? Would you consider your regulatory department effective?

Management

- How would you describe your utility's current management practices? What are they modeled after? Have they changed recently?

- For each pair of management paradigms, which one best describes water quality in your organization and why (see paper by F.G. Scott)?

  Process Control or Performance Optimization?
  Budget Administration or Resource Management?
  Command and Control or Empowered Ad-hocracy?
  Communications Channels or Communications Networks?
  Public Service or Customer Service?
  Managing Stability or Managing Change?

- Are there clear responsibilities established that govern what the water quality staff does and doesn't do? If so please provide some examples.

- What are some examples of successful management in your utility?

Personnel

- What type of evaluation system does your utility have to help identify strengths and weaknesses within your utility? Is it applied to water quality?

- Do you have any internal performance measurement systems for the water quality staff?

- How often and how do you assess your current personnel resources mix?

- In your utility, what skill sets are needed for effective production? What are the strategic "must haves"?

Communication

- Is the water quality department involved directly or indirectly in ascertaining customer attitudes toward water quality?

- How effectively does your utility communicate internally (intra-utility) and externally (public relations)? Do you have programs to refine these skills?

Decision Making

- Describe your utility's decision-making process for routine and emergency functions related to water quality?

- Have you encountered any problems in making decisions about or managing any water quality projects? If so, what was the reason and why?

Planning

- What actions does your utility take for continual improvements in water quality?

- Have you performed any water quality management benchmarking studies? What was your assessment and did it result in any organizational or management changes?

- Does your water quality department currently engage in goal setting, or does the upper management distribute the goals?

The following utilities kindly provided case study information:

- City of Ann Arbor
- American Water Works Services Company
- East Bay Municipal Utility Department
- City of Ottawa
- City of Portland

CASE STUDY #1: CITY OF ANN ARBOR

Organization

There are four divisions in the Water Utility Department all of which have some involvement in water quality:

- *Customer Service Division*
- *Wastewater Treatment Division*
- *Drinking Water Treatment Division-including distribution system water quality, pumping and source protection*
- *Field Service Division*

The Water Treatment Superintendent has a primary role in overseeing all budget and project issues related to drinking water quality. The Water Quality Manager (WQM) is a direct report to the Water Treatment Division Superintendent.

The WQM has direct involvement in:

- *Regulatory compliance*
- *Reporting functions*
- *Source water protection*
- *Distribution water quality and customer inquiries/problems*
- *Laboratory supervision*
- *Treatment optimization*
- *Laboratory research*
- *Engineering projects/engineering supervision*
- *Personnel issues*
- *Management*
- *Budgets*
- *Long-term capital planning*
- *Customer complaints*

At the moment there is no relational database system in place for the collection of lab data – Ann Arbor uses spreadsheets for data collection and reporting. The utility organizes all of its data in QuatroPRO and is slowly migrating it into Excel spreadsheets. There have been problems with operational data entry, such as several instances of gaps in the data. In addition, there have been some documented regulatory violations within the last 6 months.

Management

To facilitate an on-going transition of Management from a hierarchical system to a collaborative model, the utility has developed a mission statement and strategic plan for the future. Multiple groups were involved in the tailored collaboration of the Strategic Planning for a Utility Plan. The City of Ann Arbor does not have clear written guidelines (roles and responsibilities) for the water quality staff, but does have job descriptions.

Personnel

There is a citywide standard evaluation form used throughout the utility, however, there is not a specific water quality evaluation system. The personnel evaluation system is in need of major revisions since they are not consistently performed throughout the utility.

Decision-Making

The decision-making process is formalized in some emergency situations. Ann Arbor has extensive written emergency action plans. Typically the flow of events is as follows:

- Water Quality Manager or assistant superintendent or superintendent is alerted by Operational staff;
- WQM/Assistant Superintendent/superintendent informed; others as specified by the plan;
- State Regulatory Department is notified;
- BOIL WATER or other notice if required;
- Collection of data;
- Investigation by the State and utility working in partnership to determine what went wrong and how to correct/prevent; and
- Begin to collaboratively work on appropriate language for consumer reports.

Planning

Currently, there are no existing written programs to help refine personnel or technical skills.

Communications

Regular communication with the public is done through the web site, news articles, the CCR, City Council meetings, tours, presentations, project/neighborhood meetings. There is no official survey done to collect customer concerns. The last survey was in the mid 1980's and another is planned in the next couple of years.

Filtration Goals:

1. To prevent waterborne disease
2. To minimize disinfection byproducts and precursors
3. To minimize operational costs (pumping, on versus off peak usage)
4. To comply with state and federal regulations
5. To meet voluntary Partnership for Safe Water (PSW) goals
6. To provide aesthetically pleasing water through back up/additional taste and odor control

These are expressed as specific numerical goals.

Organization

The City of Ann Arbor is currently undergoing various organizational changes that are scheduled for implementation by June 2002. There are four divisions in the Water Utility Department all of which have some involvement in water quality:

- *Customer Service Division*
- *Wastewater Treatment Division*
- *Drinking Water Treatment Division-including distribution system water quality, pumping and source protection*
- *Field Service Division*

The Water Treatment Superintendent has a primary role in overseeing all budget and project issues related to drinking water quality. Figure B.1 below summarizes the services affiliated with the Water Treatment Division.

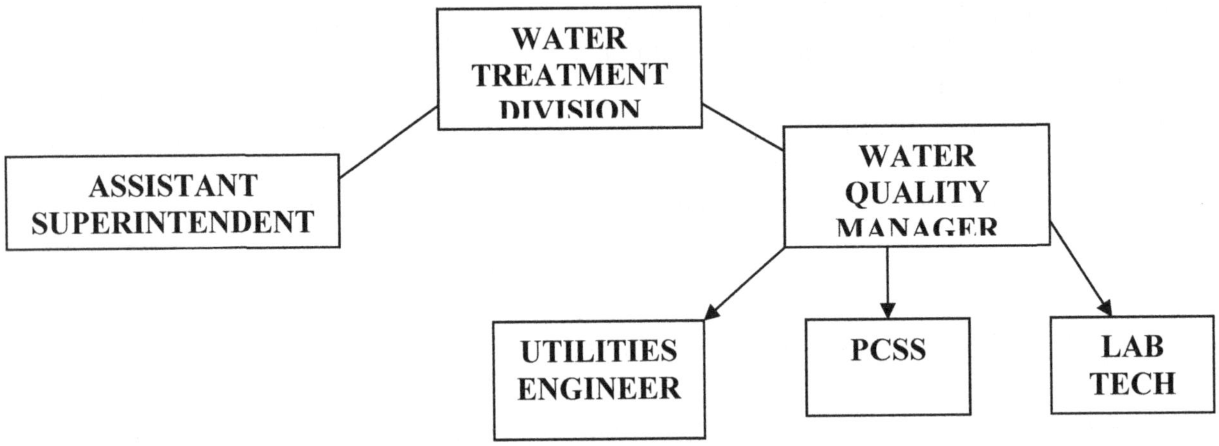

Figure B.1 Water Treatment Division Organizational Chart

The Water Quality Manager (WQM) is a direct report to the Water Treatment Division Superintendent. The WQM has direct involvement in:

- *Regulatory compliance*
- *Reporting functions*
- *Source water protection*
- *Distribution water quality and customer inquiries/problems*
- *Laboratory supervision*
- *Treatment optimization*
- *Laboratory research*
- *Engineering projects/engineering supervision*
- *Personnel issues*
- *Management*
- *Budgets*
- *Long-term capital planning*
- *Customer complaints*

The Water Quality team handles most incoming customer complaints, however, there is not one specific person assigned to public relations questions. Customer questions are initially directed to the WQM or assistant superintendent and then funneled through to laboratory staff or operations as needed. The utility has an on-site laboratory that conducts its own process control, regulatory compliance, distribution system water quality and treatment optimization analyses, which greatly reduces outsourcing costs.

The WQM also oversees a process control systems specialist and a utilities engineer. The Process Control System Specialist oversees the development and performance of the SCADA system. The utility engineer manages capital improvement projects and provides technical assistance to the various sections within the water treatment division.

For the City of Ann Arbor there are various driving factors that initiated the organizational change. Foremost was the concern that the teams within the organization were becoming isolated. The Water Quality team was consistently being overlooked and there was not an adequate communications system to voice a concern. Recently, the communication network within the divisions began opening up, but there are still communication issues. The Water Quality team meets every month. The lab meets monthly to coordinate and discuss its own projects/assignments.

At the moment there is no relational database system in place for the collection of lab data – Ann Arbor uses spreadsheets for data collection and reporting. The utility organizes all of its data in QuatroPRO and is slowly migrating it into Excel spreadsheets. There have been problems with operational data entry, such as several instances of gaps in the data. In addition, there have been some documented regulatory violations within the last 6 months.

City of Ann Arbor Filtration Goals:

1. To prevent waterborne disease
2. To minimize disinfection byproducts and precursors
3. To minimize operational costs (pumping, on versus off peak usage)
4. To comply with state and federal regulations
5. To meet voluntary Partnership for Safe Water (PSW) goals
6. To provide aesthetically pleasing water through back up/additional taste and odor control

Filtration Performance Objectives (Table B.2):

Clear wells:

1. Turbidity <0.1 NTU 100% of the time (PSW)
2. Particle counts <20/ml 100% of time
3. TTHM and HAA < 10 μg/l
4. TOC < 2 mg/l
5. 100% aldehyde removal

Filter start up spikes

1. Turbidity not to exceed 0.3 NTU for the first hour of operation (PSW)
2. Preferably turbidity not to exceed 0.1 NTU after first 15 minutes of operation (PSW)
3. Particle counts not to exceed 100/ml for first 2 hours of operation

Individual filters:

1. Turbidity for each filter < 0.2 NTU (excluding start up spike)
2. Particle counts for each filter < 40/ml (excluding start up spike)
3. Maximize run time (96 hours)
4. Minimize backwash water usage and recycle volume (backwash to 95% transmittance, this is estimated at 8 minutes in the winter and 12 minutes in the summer)

Regulations: (Current)

Clear well turbidity < 0.3 NTU, 95% of time per month
Clear well turbidity never to exceed 0.5 NTU
Individual filters < 0.3 NTU, 95% of time (15 minute readings)
Individual filters not > 1 NTU for 2 consecutive readings
Individual filters not >0.5 NTU for 2 consecutive readings (after first 4 hours of filter run)

Filtration Operations

Rate:

- Operate at <2 gpm/ft^2 (1.3 MGD); maximum allowed is 3 gpm/ft^2 (2MGD)
- Avoid rate changes on individual filters while in service, either up or down
- Bring filters on or off line to adjust for plant treatment rate; plan ahead

Run length:

- Maintain headloss < 60 inches
- ≤ 96 hours standard run

Backwash:

- Turbidity level: > 95% transmittance at end of backwash
- Standard backwash: 1 minute surface wash, 8-12 minute backwash (gradually increase the rate); adjust backwash times based on percent transmittance
- Place a new filter in service when removing one from service
- Any filter that has been out of service for 24 hours must be rewashed before it is returned to service. It is not necessary that this be done at 24 hours, but rather in anticipation of when a filter will be needed.
- When returning a filter to service, it is desirable that filters that are off for long hours be returned to service first (minimizes algae growth)
- Backwashing uses energy; avoid peak energy charges by backwashing filters between 7 PM and 11 AM. Consult with the Operator III before backwashing on-peak (11 AM to 7 PM). Do not rewash filters until they are needed (or algae/other problems develop)

Start up:

- Wait one hour after backwash (and rewashing) before starting filter
- Apply alum at filter start up

Monitoring:

- Turbidity – on-line
- Particle counts – filters #6 and #19 on line
- Percent ash – 2 filters/6 months
- Effective size and uniformity coefficient – 2 filters per 6 months
- Iodine number – 2 filters per 6 months
- Apparent density – 2 filters per 6 months
- Abrasion number – 2 filters per 6 months
- Aldehyde removal – monthly

- ♦ BDOC removal – monthly
- ♦ Backwash water transmittance – filters #6 & #19 on-line

Table B.2: City of Ann Arbor Operational and Water Quality Goals

Summer

	Primary Basin	**Secondary Basin**	**Ozone**	**Wet Well**	**Filters**	**CW**	**Reservoir**	**Dist. Syst.**
pH	11.0 -11.3	≈ 10	8.0	9.3	9.3	9.3	9.3	
sludge %	8-9%	≤ 3%						
[OH]		≈ 0						
[polymer]		0.6 mg/l						
turbidity		< 5 NTU			<0.2 NTU	<0.1 NTU	<0.1 NTU	<0.5 NTU
Average [O_3] resid			0.1 mg/l 1st cell					
% CT			150-200 %					
[PO_4]				0.95 mg/l as SHMP				
filter run					96 hours			
Backwash					1 min surface. 8 min regular.			
Cl_2:NH_3-N						4.75-5.00 : 1		
Alkalinity							>30 mg/l	
TH							120 mg/l	
Fluoride							1.0 mg/l	
NH_2Cl							3.0 mg/l	>2.0 mg/l
T&O							0	0
Nitrite (μg/L)								<100
HPC								<500 cfu/ml

(Continued)

Winter

	Primary Basin	Secondary Basin	Ozone	Wet Well	Filters	CW	Reservoir	Dist. Syst.
pH	11.3 -11.5	≈ 10	8.0	9.3	9.3	9.3	9.3	
sludge %	8-9%	≤ 3%						
[OH]		≈ 0						
[polymer]		0.6 mg/l						
turbidity		< 5 NTU			<0.2 NTU	<0.1 NTU	<0.1 NTU	<0.5 NTU
Average $[O_3]$ resid			0.1 mg/l 1st cell					
% CT			150-200 %					
$[PO_4]$				0.95 mg/l as SHMP				
filter run					96 hours			
Backwash					1 min surface. 8 min regular.			
Cl_2:NH_3-N						4.75-5.00 : 1		
Alkalinity							>30 mg/l	
TH							140 mg/l	
Fluoride							1.0 mg/l	
NH_2Cl							3.0 mg/l	>2.0 mg/l
T&O							0	0
Nitrite								<100 µg/l
HPC								<500 cfu/ml

Notes: 1) Maximum well water in primary ≥75%. 2) Can move backwash recirculation to secondary if having hexahydrate sludge problems. 3) Add $CaCO_3$ to primary RM as needed for hexahydrate problem

CASE STUDY #2: AMERICAN WATER WORKS COMPANY

American Water Works Company (AWWCo.) is the largest publicly traded U.S. corporation devoted exclusively to the business of water. Its 6,500 associates provide water, wastewater and other related services to 15 million people in 27 U.S. states and three Canadian provinces. American Water Works Company is comprised of a number of subsidiaries and provides contract managing and operating services to 175 water and wastewater facilities in 7 states. The American Water Works Service Company, Inc. (AWWSC) provides technical and professional services to its utility subsidiaries.

Organization

The Water Quality Department is located in the AWWSC and is headed by the Vice-President of WQ, who is responsible for overseeing research, laboratory, and environmental compliance. Figure B.2 illustrates the American Water Works System Corporate Organization for the Water Quality Department.

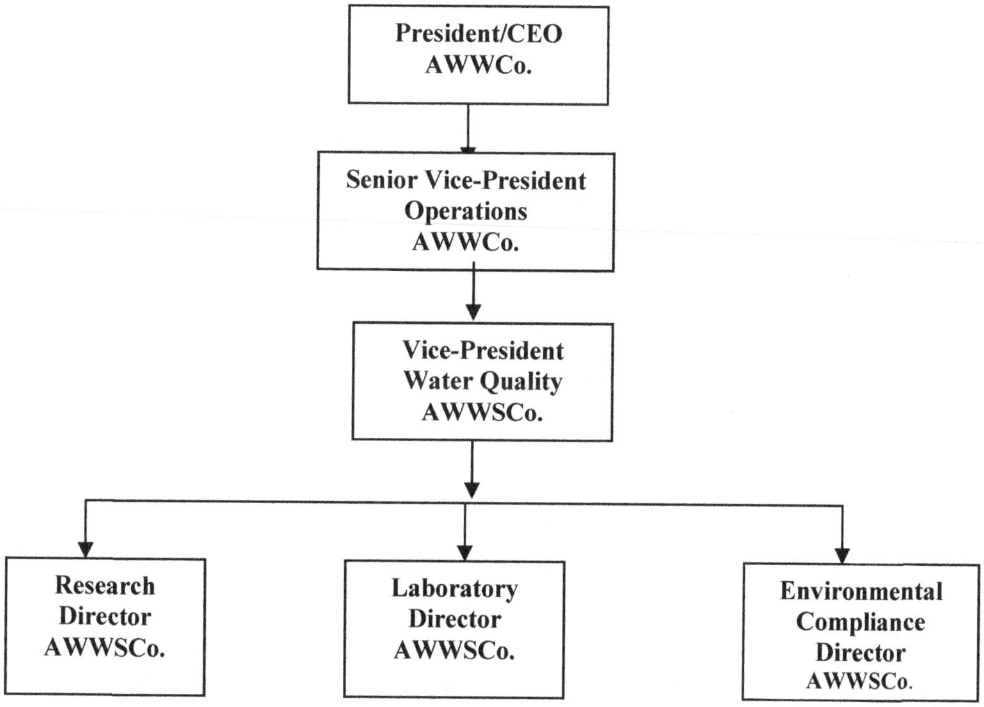

Figure B.2 American Water Works Company: Water Quality Department Corporate Organizational Chart

In addition, each utility subsidiary also has a Board or Directors, company president, vice president of operations, and a director of water quality. The director of water quality typically supervises a statewide staff of water quality specialists, analysts, and technicians. The director of water quality at the operating utility reports both to the vice president of operations at the state level, and the vice-president of water quality at the corporate level. This dual reporting provides a check and balance to ensure that each system meets water quality goals. Besides regulatory limits, AWWSC has established its own numeric and descriptive water quality goals that provides guidelines to the operating utilities. However, water quality requirements may vary from state to state, and the Water Quality Director (WQD) in each respective company is responsible for maintaining compliance with state requirements. The AWWSC functional groups in water quality are primarily responsible for setting corporate policy.

There is a central water-testing laboratory, the Quality Control and Research Laboratory, located in Belleville, Illinois. Certified in 25 states, the Belleville Lab performs compliance testing for almost all regulated contaminants for every subsidiary. By having one main lab, data on drinking water quality is generated internally at the central location and guarantees uniform quality control and assurance. The result is an extensive database system that is voluntarily shared with government agencies including the USEPA. At this time the lab is in the process of developing electronic data reports for each state so as to efficiently distribute information to the respective subsidiaries. There were no regulatory violations for any of the individual companies in 2001.

Management

The organization is a corporate business, and abides by business models for evaluation of resources and allocations of capital. The most challenging aspect is the development of a network to effectively communicate, as it is more and more difficult to gather individuals to one site for meetings. Customer Service is a major driver in the organization, and there is an incentive program directly geared to meeting compliance and reducing the number of customer complaints. To meet its "world class" service objective, there is a continuous goal to achieve 90% customer satisfaction based on a "good" or "excellent" response for an independent customer survey. AWWC believes that satisfied customers are essential to the continued success of the company, and this ideal is incorporated in the organization's Mission Statement. Managing stability within this organization is always challenging since the company is always expanding and purchasing new companies.

Personnel

An annual Performance Appraisal evaluation system with several interim evaluations and a leadership development program helps identify personnel strengths and weaknesses and keeps associates focused on the overall corporate goals. Within the water quality function, there are functional meetings both at the operating utility and corporate level. The Human Resource Department also offers management and technical development courses. The VP of Water Quality is extensively involved in the training program with the water quality staff, to prepare them for future issues. Guest speakers, experts in certain fields, are frequently invited to update the staff on many technical developments. A big emphasis is placed on valuable employees, which undoubtedly contribute to the organization's success. Some of the skill sets which were identified as "strategic must haves" are:

- Ability to communicate
- Internal and external skills
- Ability to motivate others
- Personal motivation
- Technical skills.

These skills, in that order, are essential in a highly productive employee.

Communication

All Water Quality Directors participate in a monthly conference call to coordinate water quality activities. As previously mentioned, customer attitude surveys are continuously performed and summarized on a quarterly and annual basis. In addition, project management programs are geared towards maintaining communications with the public and updating them on the progress of a various projects. Internal communication networks are currently being improved through the development of an intranet. Corporate research, environmental compliance, and laboratory activities are reported on a monthly basis. The research group regularly produces project reports and journal publications.

Decision-Making

Each subsidiary's decision-making process is collaborative. However, the Corporate Office manages decisions dealing directly with the overall business. In terms of emergency response decisions, there is an Emergency Response Center that has its own communication system to resolve problematic issues in a timely manner. There have not been problems, or lack of resources, associated with management decision-making.

Planning

The Research Program, part of the American System Water Quality Department, is staffed by 20 professionals who are constantly investigating future research issues. These issues are picked up and developed by the engineering planning group that develops 5-year and long-term plans for each individual water company. When these plans are ready to be implement, the engineering design group develops specific facility projects. In this manner water quality and engineering are working together to ensure that operating companies have the resources to meet customer expectations.

Philosophical Approach (see Figure B.3)

American believes that satisfied customers are essential to the continued success of the company. In its mission statement, American states that it is critical to constantly build increased value to customers. Our vision is to provide unsurpassed water service in order to protect public health and promote quality of life. American recognizes the critical role water service (and high quality drinking water) plays in its customer's lives. Regarding drinking water and the environment, American has held that to fulfill its commitment, it must

- Provide high quality drinking water, surpassing all standards
- Educate the customer on water-related matters
- Be environmentally conscious
- Be a leader in the industry
- Support regulations that improve the health of customers
- Conduct itself with high integrity
- Exceed customer expectations

Figure B.3 American's Water Quality Philosophy

Water Quality and Environmental Matters

American's water quality organization was established as a distinct department twenty years ago, for the expressed purpose of emphasizing the importance of drinking water quality to the company. During the intervening years, American has shared data on water quality, conducted research directed at answering regulatory questions, participated in EPA advisory groups, and conducted technical sessions specifically for the agency.

For two decades, American has conducted a formal research program using a cadre of full time, dedicated professionals on staff. Today there are twenty researchers, including microbiologists, chemists, and engineers. This research program, costing American about $3 million annually, is directed at providing answers to issues of importance to American's utility subsidiaries, including compliance with mandatory regulations. Conducting research provides American with advanced knowledge on the effectiveness of treatment options, and future capital investment needs.

American has been very active in the development of regulations at the national level. Besides providing insightful comments, based on extensive experience owning and operating water systems, American has been officially involved by having associates serve on both EPA advisory boards to the Office of Ground Water & Drinking Water (Science Advisory Board, and National Drinking Water Advisory Council), plus numerous EPA committees and working groups.

Compliance and Going Beyond

American is committed to the highest level of compliance with the spirit and letter of environmental and drinking water regulation. A component of the annual incentive compensation available to our associates requires each utility to be in full compliance with all environmental regulations.

Going beyond compliance, and recognizing that just meeting the required standards is not enough to satisfy customers, is one element of American's vision of industry leadership. American sets its own internal drinking water goals, which are a combination of regulated and unregulated contaminants, to include many unregulated substances that a consumer would find objectionable.

To attain the highest quality drinking water from its surface water treatment plants, American voluntarily joined in the EPA Partnership for Safe Water (Table B.3). The Partnership was conceived as a program to reduce the likelihood of waterborne exposure to *Cryptosporidium*. American's record of achievement in this program further demonstrates its leadership.

All of the 81 surface water treatment plants owned by American are part of the Partnership program. 55 have received the prestigious EPA Director's Award with the remaining plants in the process of obtaining the award.. Comparing these numbers to the remainder of the industry where only 3.5% of plants have entered the Partnership program, and only 31% have received the Director's award, American has more than doubled the achievement of the industry.

Table B.3 EPA Partnership for Safe Water

	American Water System	Remaining USA Systems	USA
Community Water Systems using Surface Water	81	11,322	11,403
Partnership for Safe Water participants	81 (100%)	320 (2.8%)	401 (3.5 %)
Systems receiving the Director's Award	55 (68%)	98 (31%)	153 (38%)

Satisfied customers and communities are the foundation of a successful water utility. American utility subsidiaries conduct quarterly satisfaction surveys of customers and public officials in the communities we serve. American has taken actions based on those surveys with the goal to achieve a 90% satisfaction rate for water quality and an 85% overall level of satisfaction – a world-class achievement. Currently water quality satisfaction rate (a rating of good or excellent) is 75% and community performance rate is 84%. We are proud of those numbers, but are diligently striving to increase them.

Over 20 years ago, American established a policy of chlorinating every water supply, including ground waters. While this practice was controversial, today it is the basis of the pending Ground Water Rule – an accepted practice as we learn more about microbes in aquifers.

Communication to customers and public officials has been a steadfast practice of the American water System. For 15 years, our utility subsidiaries have voluntarily issued annual reports on drinking water quality. So when the 1996 Safe Drinking Water Act required Consumer Confidence Reports, it was a verification that American's practice was ahead of its time.

Another example is the voluntary issuance of a Notice to Immunocompromised Customers. Being concerned about waterborne disease outbreaks of Cryptosporidiosis, which creates serious health problems for immunocompromised individual, American launched a voluntary communication initiative a few years ago. A special attention-getting brochure was developed, with the help of an expert panel assembled including medical doctors, public health officials, communications experts and AIDS groups, and mailed to all customers. It advises of the seriousness of contracting this disease, the possible routes of cryptosporidiosis, and offers tips to avoid exposure to the organism along with a recommendation to seek medical advice if the customer has a severely suppressed immune system. Issuing this notice was a bold decision for American, but one that was needed for protection of all customers. American utility subsidiaries send this brochure to all customers on an annual basis.

Summary Points

- Outstanding record of compliance
- Achieves drinking water quality surpassing standards
- Environmentally conscious
- Extensively involved in regulatory development
- Supportive of additional health protection
- Very sensitive to customer satisfaction
- Performs water testing centrally, doing considerably more than required
- Conducts research related to regulations, and shares information with EPA & State agencies
- Provides laboratory test data to EPA for use in future regulatory decisions
- Provides advice to EPA through committees & working groups
- Educates customers on water-related matters

CASE STUDY #3: EAST BAY MUNICIPAL UTILITY DISTRICT

The East Bay Municipal Utility District (EBMUD) is located in Oakland, California. EBMUD supplies water and provides wastewater treatment for parts of Alameda and Contra Costa counties on the eastern side of San Francisco Bay in northern California. A population of approximately 1.2 million people is served at 192 MGD.

Organization

A seven member Board of Directors determines overall policies, which are then implemented under the direction of the General Manager. Each Board member serves on one or more standing Board committees. The committees review and make recommendations with regard to matters of major interest, prior to discussion by the full Board at a regular meeting. Standing committees of the Board include Planning, Legislative/Human Resources, and Finance/Administration.

The Operations and Management (O&M) Department is responsible for water treatment and distribution and some source water issues. The Department is the area primarily responsible for overseeing all issues related to water quality. The Operations Division is then comprised of Operations Engineering, OP/NET Support, Water Distribution, System Water Quality, and Water Treatment.

The Operations Engineering Department is comprised of 5 research engineers whose primary responsibility is to conduct research for source, treatment and distribution optimization. The Water Distribution department is responsible for all quantitative functions and controls distribution flow. The System Water Quality Department has a primary role of interfacing with customer complaints, monitoring, and distribution system sampling, flushing and chlorination. Water Treatment is accountable for all day-to-day operations of 6 water treatment plants. The OP/NET Support Department has a primary role in the operation and networking systems such as the SCADA system. The Water Quality Office has the responsibility of oversight of all water quality issues within the District.

Water Quality samples are analyzed in an in-house laboratory. The System Water Quality and Plant operation staff collects the samples, the samples are then entered in the data system, and the lab sends the data to the program manager. Very few sample analyses are outsourced. Some of the outsourced samples are asbestos and other unregulated contaminants. A recently completed assessment identified the need for a water quality database system for storing all water quality data. To complement laboratory information management system (LIMS) the organization has not had any violations within the last 6 months.

Management

The organization's management practices are directed at performance optimization and resource management. This management practice is described as very efficient by the organization, as there have not been any problems in meeting any of their goals. .

One example of successful management is the development of an emerging contaminants protocol,. Customers were aware of high levels of MTBE present in their drinking water supplies, and expressed their concern to the utility. By following a set management protocol, the Utility was able to develop goals and then implement an action plan to remedy the problem. The implementation of an action plan, including operational changes and capital investments were made, facilitating a resolution in a short period of time.

Personnel

The utility utilizes an annual and interim performance plan system tied to compensation to identify personnel strengths and weaknesses, which helps in ascertaining which employees are meeting their goals. Water Quality goals are distributed by the WQ Committee, these goals are attentive to all water quality issues and are not specific to any one department.

There are certain staff qualities that an organization will benefit from and contribute to the overall efficiency. Some of the mentioned strategic must-haves include:

- *Academic and technical qualifications,*
- *Motivation, and*
- *Versatility* (ability to move across departments within the organization).

Communication

At the moment, there are internal newsletters that are distributed throughout the utility to update everyone on the latest issues. The monthly Water Quality Committee meeting facilitates performance monitoring, emerging issues, regulatory development, goal setting and strategic direction. In addition to these updates, there are programs in effect, such as the Engineers Forum and the Brown Bag Forum, which promote internal communications. External communications include participation in school fairs, school programs and brochures beyond the Consumer Confidence Reports. The Water Quality Manager reports semi-annually to the Board of Directors on Water Quality Issues. This is a public presentation. There are further steps being taken to promote customer education, one of those being communicating information over the Internet.

Decision-Making

Routine decisions are made within the functional units. Major projects follow a more formal decision process. For emergency response, the decision-making process is documented, well-developed, and routinely exercised. . Emergency decisions consist of various departments within the district. All of the individuals that are pertinent to the situation are relocated to an Emergency Operations Center, which has its own communications system. This type of management has proven to be very effective.

Planning

The utility is a strong goal setting organization. The whole O & M team will participate in a two-day retreat to evaluate goals and plan for the future. In addition, an in-depth master plan has been developed in the past two years to address Treatment and Transmission. A Watershed Master Plan was developed in the late 1990s. .

This organization has identified the major priorities for management:

- *Budget*
- *Water Quality Violations*
- *Water Quality Commitment*
- *Short-term Water Quality Goals*

In addition to the master plans, there is a semi-annual report that is developed by the Water Quality Department and sent to the Board of Directors in order to keep them updated with all water quality related issues. The organization has also participated in benchmarking studies.

Water Quality Goals

Water quality goals can either be expressed as treatment performance criteria or desired contaminant concentrations. They are intended to meet actual or anticipated regulatory requirements, customer expectations, and technical/financial feasibility criteria. It is important that goals be documented and be quantified. Over the past few years, EBMUD has set two levels of goals: Annual Goals and Long Term Goals. In recent deliberations, an overriding principle in goal setting has been established by the District to have water quality goals that will consistently deliver water quality (for all parameters) that is in the top 10% in the state.

Annual Goals (Table B.4) are achievable with operational changes and process optimization. Once a month, the District's Water Quality Committee that includes members from operations and maintenance, engineering, source water treatment, and distribution, reviews the progress on meeting the Annual Goals.

Table B.4 Annual Water Quality Goals

Parameter	Goal*	Regulation	Compliance Point
THM (Trihalomethanes) System Wide Running Annual Average	40 ppb	80 ppb	Distribution System
THM (Trihalomethanes) Locational Running Annual Average	60 ppb	Future D/DBP Rule	Distribution System
MTBE (Methyl tert Butyl Ether)	0.5 ppb	5.0 ppb secondary 13 ppb primary	Raw Water Intake or Plant Effluent
Customer Complaints	<50/month	N/A	N/A
Customer Complaints	0 > 50 customers	N/A	N/A
Turbidity (Daily Maximum)	0.1 NTU	0.2 NTU 95% of the time	Combined Plant Effluent

Parameter	Goal*	Regulation	Compliance Point
Turbidity (Spikes in filter effluents)	0.3 NTU 0.1 NTU in 15 min.	1.0 NTU	Individual Filter Effluents
pH (Entering the distribution system)	pH > 8.5	Lead & Copper pH > 8.0	Combined Plant Effluent
TOTAL COLIFORMS (as measured in TCR samples)	0.5 % coliform positives/month	5% positives/ month	At TCR taps in the distribution system
Heterotrophic Plate Count (HPC) in Distribution taps	50 colonies/ ml	500 colonies /ml	At TCR taps in the distribution system
HAA (Haloacetic Acids) System Wide Running Average	30 ppb	60 ppb (proposed)	Distribution System
HAA (Haloacetic Acids) Locational Running Annual Average	45 ppb	Future D/DBP Regulation	Distribution System
Bromate System Wide Running Annual Average	5 ppb	10 ppb D/DBP Stage I	Combined Plant Effluent
NDMA (N-nitrosodimethylamine) Locational Running Annual Average	5 ppt Plant Specific Running Annual Average	CA Action Level 10 ppt System Wide Running Annual Average	Distribution System
Chlorate	0.6 ppm	0.8 ppm CA Action Level	Distribution System

ppm = parts per million = milligram/liter NTU = Nephelometric Turbidity Unit
ppb = parts per billion = micrograms/liter
ppt = parts per trillion = nanograms/liter

The Water Quality Goals establish specific numeric goals for water treatment plant effluents and the water distribution system to ensure all customers get high quality water. Goals address both potential health risk and aesthetics. Regulations represent the minimum goal level. Customer expectations, economic considerations, and technical feasibility and competing regulations are considered when goals are developed.

For those goals not being met, all involved Departments will strive to meet these goals through operational optimization. Specific actions to meet these goals will be documented and adopted at the beginning of each calendar year by the Departments.

Long-Term Goals are set to incorporate current and anticipated regulations and customer expectations as well as to address trends that are foreseen over the next 10 to 20 years. EBMUD's Long-Term Goals were set at a workshop attended by staff and national water quality experts in 1999 and refined in the Master Plan Process. Each goal is supported by a set of criteria to determine how the goals are met. The Long-Term Goals are used to guide capital treatment upgrade programs and are used by management to review program performance.

Goal 1. To provide all customers comparable water quality that meets or exceeds customer expectations with regard to contaminants with significant public health impact and aesthetic qualities.

Goal 2. To seek the best available water source and protect them from potential degradation.

Goal 3. To seek cost effective treatment technologies that produce high quality water, minimize chemical usage, are robust and flexible and can be efficiently operated and maintained.

Goal 4. To minimize degradation of water quality through the distribution system.

Goal 5. To minimize adverse environmental impacts from operations

Action Plan

Once goals have been established, the next step is to develop an Action Plan to meet the goals. The goals may require operational changes or capital projects. As an example, the MTBE Action Plan included both capital and operating initiatives:

- Replacement of all rental and District-owned high emission 2-cycle engines
- Increased monitoring to better determine the fate and transport of MTBE in reservoirs
- Public outreach to keep drinking water and recreational customers informed
- A ban on use of all high-emission 2-cycle engines on one reservoir
- Participation in statewide efforts to reduce or eliminate the use of MTBE in gasoline
- Optimization of boat fueling practices at reservoirs.

Water Quality Data Base

A water quality database, including regulatory and operational monitoring parameters from field, on-line and laboratory sources in context with environmental and treatment parameters, is essential to monitor and document the ongoing water quality and the effect of capital and operating changes made. Currently numerous water quality database existing in different programs are distributed throughout the District.

CASE STUDY #4: CITY OF OTTAWA

The City of Ottawa, located in Ontario, Canada, currently serves approximately 800,000 people. It has a combined treatment plant capacity of 200 million gallons per day (MGD), however it typically only uses approximately 55% of it capacity. In this province, individual utilities are composed of various cities and organized together under one "central mega-city concept." The City of Ottawa recently underwent such a consolidation to improve operating efficiencies.

Organization

The General Manager is responsible for all utility services including water and wastewater treatment, landfills, environmental programs, as well as five ground water systems and communal wells currently in operation. The organization's Drinking Water Services Department under the Utility Services Director which reports directly to the Transportation, Utilities, and Public Works General Manager. Other departments such as Wastewater and Drainage Services,

Solid Waster Services, Environmental Programs and Technical Support, and Customer Services and Operational Support all have similar structure.

Water Quality has resided under the Production Branch for over 70 years. The Production Branch is primarily responsible for water treatment plant operations, water quality technology, remote pumping and storage, SCADA, water quality processes, laboratory function and operations support.

The Water Quality Department (WQD), is involved in all water quality monitoring and reports. WQD is in the Production Division and is seen as an "oasis" of technical knowledge and therefore takes on the responsibility of being the "problem solver" for the utility. Furthermore, the WQD is directly involved in plant operator training, source water protection, emergency response, and well system development.

In terms of a database management system, the organization is currently in the process of developing a more efficient data management systems than the ones utilized to date. Data collection is typically entered into Excel spreadsheets by laboratory staff. This utility processes a total of 55 data reports through an Oracle data management system and the Laboratory Information Management System (LIMS). The City of Ottawa has not had any regulatory violations in the past 6 months.

Management

In August of 1999, a plan for major reorganization of the utility was developed, but has not yet been implemented. The utility modeled the new organization after the Production Branch so that each unit would become its own operating facility. The plan has not been implemented primarily due to a lack of a communications network. The utility has identified the management decision-making process as a hierarchical process, one that does not foster collaborative decisions. It is currently taking the appropriate measures to remedy this. Another area of concern is the lack of established responsibilities identified for each role in water quality.

The research area has been very successful and has received the appropriate level of support from the City Manager. Currently, there are 23 ongoing projects, which will prove beneficial in future plant optimization.

Note: Ottawa has one of the most progressive research programs in the continent and this should be emphasized.

Personnel

The City has a performance review process. An annual review process has been developed, however, resources do not facilitate the consistency of this review process. Currently, reviews are performed on an Ad-hoc basis, there are no set schedules and reviews occur only when needed. The utility has identified several strategic skill sets considered "must haves" for effective production. Some of these include:

- ***Personal Skills*** - Staff members that are able to work as a team;
- ***Problem solving skill*** - Being able to use common sense to prioritize and analyze problems;
- ***Knowledge*** - Having the technical knowledge required in running a plant.

Communication

The Water Quality Department is directly involved in external communications with the public. The WQD handles all incoming customer complaint calls and gives advice over the phone to the customer. If an issue cannot be resolved the call is forwarded to an operator for further investigation. There are no documented guidelines for this procedure and annual customer surveys are not conducted. Internal communication networks do not exist, as there are not any scheduled internal staff meetings.

Decision-Making

The utilities decision-making process for routine and emergency response is informal. There are no documented procedures. Staff resources have hindered some project decision-making, as there is not enough staff to execute projects.

Planning

City of Ottawa has developed short-term and long-term strategic plans. Long-term plans extend as far as 20 years in advance; these plans call for a devotion of resources by the utility for infrastructure rehabilitation. In addition short-term plans have also been initiated, which include a 2-yr. plan for facility expansions and optimization. Regular supervisor/manager meetings to develop area specific plans for the department have not been conducted.

CASE STUDY #5: PORTLAND WATER BUREAU

The Portland, Oregon Water Bureau engages in a distributed approach to water quality management. The Portland Water Bureau serves a population of approximately 800,000 including about 60% retail and 40% wholesale customers. The supply system includes surface water and groundwater sources and has a total capacity of about 300 MGD. Average annual demand is about 110 MGD. There is no single specific section within the Operations and Management Group that has exclusive responsibility for water quality functions.

Organization

Within the Portland Water Bureau, the Operations and Management Group is responsible for overseeing major water quality functions within the utility. Water quality roles are embedded in various sections within this organization. The primary components associated with water quality include: laboratory services, regulatory compliance, water quality inspections, and water treatment. Other water quality-related functions in the group include flushing, system operations, watershed management, and control systems (SCADA and instrumentation). In addition, some water quality related functions are also part of the Engineering, Resource Protection, and Public Information Groups within the Bureau.

Figure B.4 below summarizes the services affiliated with the Operations and Maintenance Group of the Portland Bureau of Water Works:

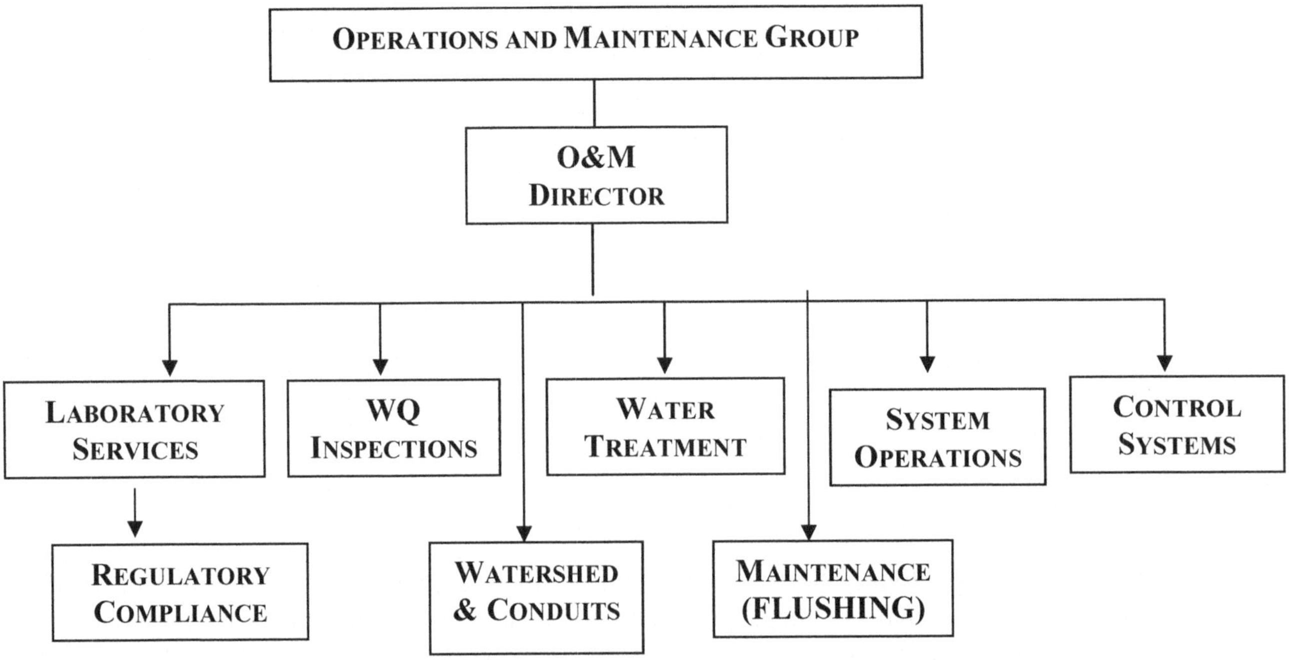

Figure B.4 Portland Bureau of Water Works: Operations and Maintenance Group Organizational Chart

Each section under the O & M Group (i.e. lab services, water quality inspection, water treatment) has a manager who reports directly to the O&M Director. The Laboratory Services section is categorized as a typical municipal drinking water lab. Its primary focus is to conduct distribution system bacteriological analysis and physical analysis of samples from the supply sources. Several chemical analyses, such as VOCs and THMs, are done in-house to minimize outsourcing expenses; however, many analyses are contracted out. The lab manager is responsible for coordination of all lab functions and lab personnel issues. The Lab Manager has technical qualifications that includes a BS and graduate studies in Chemistry, and has worked in an academic lab for 10 years. The Lab Manager also has the opportunity to enhance his/her skills sets by participating in various management courses offered by the city.

The Regulatory Compliance section staff is responsible for the development of monitoring schedules and ensuring that schedules meet applicable regulatory sampling requirements. Once samples are analyzed, this section is responsible for consolidating data from multiple sources (LIMS, SCADA, manual reports) to produce all required reports (i.e. monthly surface water reports, TCR reports, distribution system reports). Furthermore, the Regulatory Compliance section must make certain they keep up to date on emerging regulations.

Figure B.5 illustrates the flow of information between the Regulatory Compliance Group and Laboratory Services.

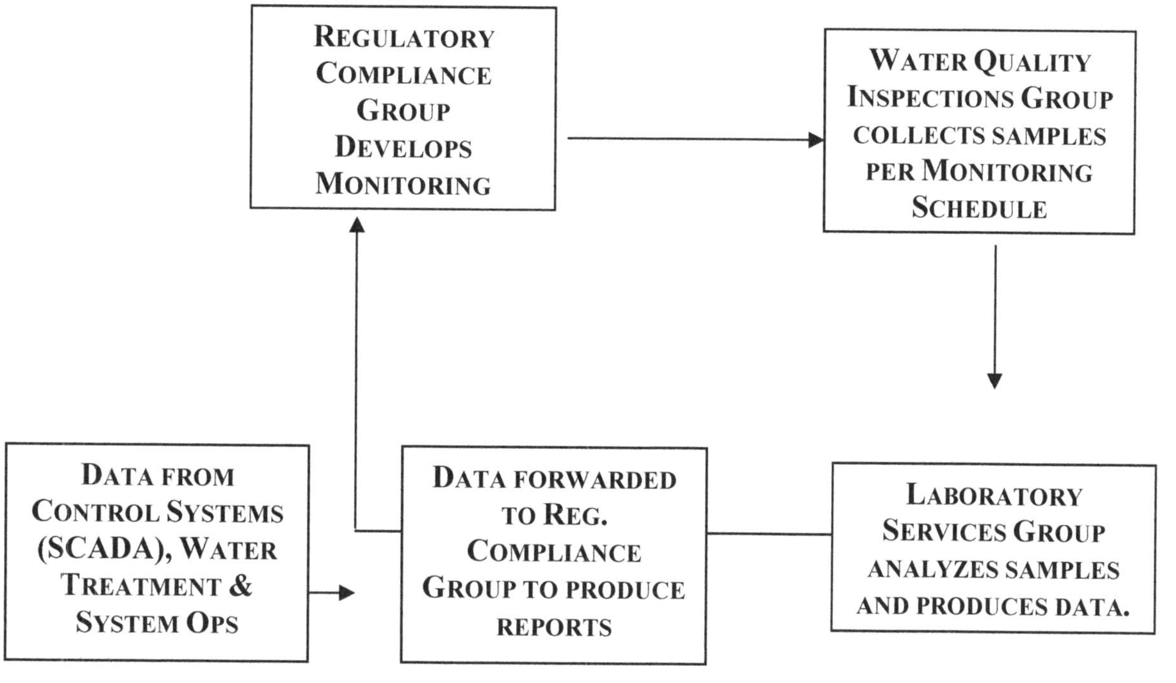

Figure B.5 Relationship between Regulatory Compliance and Laboratory Services

The Water Quality Inspections Group has 4 major roles:

- ***Cross-Connection Control*** **-** These duties include notification and record keeping, facility and device inspection, plan review and establishing requirements.
- ***Distribution Sampling*** **-** This includes collection of about 400 bacteriological samples per month. A total of 2 WQ Inspectors collect the samples and input the data via a barcoded samples into a LIMS. Additional samples for nitrification monitoring, disinfection process control and corrosion control compliance are collected periodically.
- ***Supply Sampling*** **-** These duties include source water sampling of the watershed and groundwater well field samples.
- ***Customer Complaints*** **-** These duties include customer interface in response to questions, concerns and complaints relating to water quality.

Customer concerns are a very important component in Portland's water quality program. Figure B.6 illustrates how this organization addresses these concerns in a formalized manner.

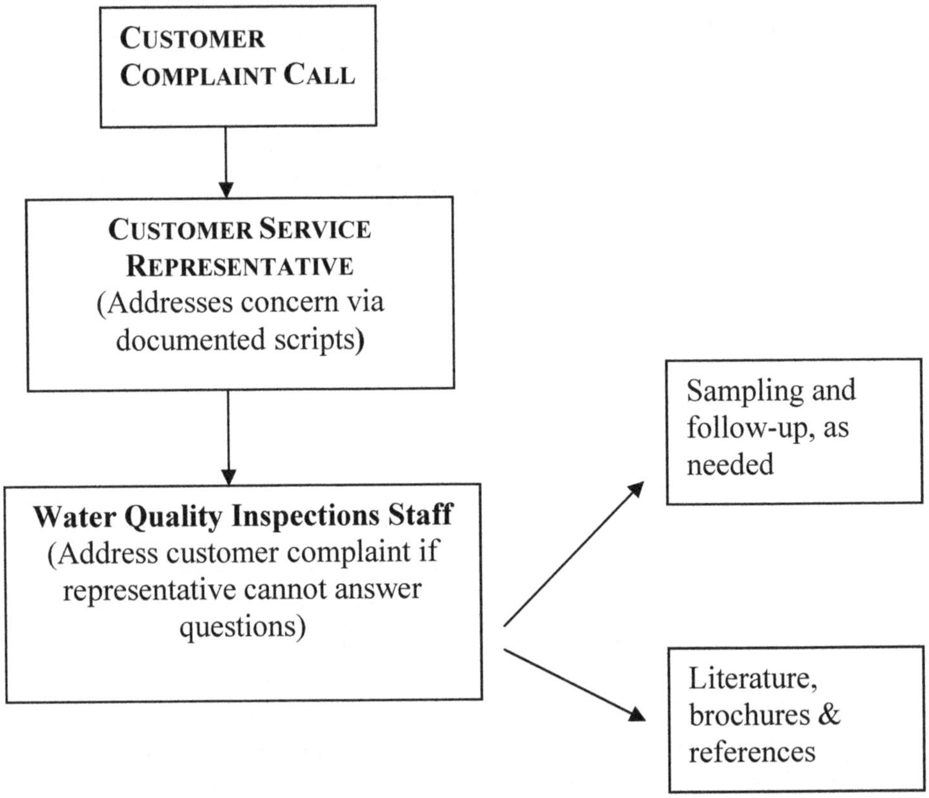

Figure B.6 Customer Complaint Flow Chart

This initial points of contact for customer complaints are the customer service representatives. Frequently, complaints can be addressed through the use of documented scripts developed for the service representatives to address common complaints. If the complaint can not be adequately addressed through this preliminary step, it s referred to the water quality inspections staff, where follow-up actions can be taken, ranging from providing the customer with advice or informational brochures, to water sampling and follow-up as needed.

The Water Treatment section is accountable for all surface water supplies. This includes disinfection treatment and pH adjustment. A relatively small staff, totaling 10 operators and 1 manager operate this section on a 24-hour basis. Changes in water treatment are based on input from the Control Systems Section. SCADA systems record all reservoir turbidity, temperature, pH, and chlorine residual levels. All necessary adjustments can be made remotely. Laboratory and Regulatory Compliance sections collaborate with the Water Treatment and System Operations sections to establish operational targets for water treatment.

Each section previously described conducts its own internal staff meetings every other week. In addition, every other week the section managers meet in an "Operations Coordination" meeting (Ops-Coord) designed to examine every aspect of the system in order to identify major issues and to prepare for future issues. It is a very interactive organization with an efficient communication system. The Portland Water Bureau indicated that the "Ops-Coord" process allows for efficient and collaborative analysis of water quality issues by providing a diverse team-based evaluation of complex issues. This approach also lends itself to identification of opportunities for ongoing improvements and by eliminating supervisor layers.

The Portland Water Bureau identified that areas requiring further improvements include flushing, and coordination with Engineering and Resource Protection groups.

Management

The organization recently underwent a financial and resource crisis resulting from problems with a new utility billing system. These problems resulted in a critical shortfall in incoming cash-flow and dramatically increased expenses associated with support of the billing system. All available resources were redirected to the billing system department and Operations staff was on loan to help alleviate billing issues. While the core business of providing high quality drinking water was not compromised, this condition has resulted in a notable reduction of resources in all groups and has brought about deferrals of maintenance and various system improvements.

Currently, the organization does not have documented roles and responsibilities for its staff other than generic civil-service job descriptions. It is generally assumed that each employee has a clear sense of what his/her contribution to the organization should be.

Examples of successful management within the utility include the development of a Maintenance Pilot Program. The plan incorporated the collaborative efforts of all the groups involved. The program demonstrated the power of the cross functional approach. Generally six individuals from two groups within the Bureau (Operations and Construction) completed replacement of all meters, performed meter-box maintenance, exercised existing gates and performed preventative maintenance on all hydrants in a 32-block area of the city. Innovative scheduling and staffing allowed this crew to complete its work in half the time and with half the resources of what it would normally take.

Personnel

A formal system to evaluate the utility's strengths and weaknesses does not exist. A standardized personnel evaluation system is in place throughout the city; however, all employees do not consistently do these evaluations. There is an evaluation of resources as part of the development on the utility's annual budget. The City Auditor conducts and annual review of service and performance measures that includes comparisons of selected factors (including water quality measures) to other comparable-sized utilities as well as tracking these changes over time.

The utility staffing generally reflects a highly technical and engineering-oriented management staff. In general, senior managers come from an engineering background, with demonstrated project management skills considered as "must haves" for effective production.

Communication

The Operations and Maintenance Group is actively involved in monitoring customer attitudes. Every year a survey is distributed to the customers. The survey does not specifically target water quality functions, instead it asks about the utility as a whole

Decision-Making

The organization's routine decision-making is a collaborative effort between the staff and managers of the various sections within the O&M Group. The O&M Director ultimately decides what changes need to be made regarding treatment and operational issues based on recommendations from section managers and staff. While this collaborative style derives benefits from having input and analysis by multiple disciplines, lines of communication and authority have a tendency to become clouded at times.

Emergency decision-making is described as extremely efficient. The utility has fully adopted the Incident Command System (ICS) for responding to emergencies. This "top down" hierarchal style of communication and management allows for very direct lines of authority with established procedures for documentation and information dissemination. The utility has conducted numerous drills and exercises, as well as used the ICS in several system-wide water quality emergencies in the recent past. There is now almost an automatic shift towards this militaristic decision style in times of crisis. Once an Incident Commander initiates a response, the appropriate staff is transported to a 3,000 square foot command center and there is a shift to the incident command system. Typically any problems associated with decision-making in this utility are based on lack of time, budget, or personnel resources.

Planning

The organization implements several tactical short and long term plans to prepare for the future. A 10-year horizon treatment process plan is continually being developed. In addition, the Bureau maintains a 10-year Capital Improvement Program, which includes numerous large water quality-related projects (i.e., 250 MGD filtration plant, replacement and covering of open reservoirs, 90 MGD groundwater treatment improvements). Daily operation improvements are targeted in a 2-week plan. To address the gaps between the long-term and short-term plans a seasonal plan, 6-months to 2-years, is developed.

Summary

The case studies provide a representative sample of 'real world' water quality organizations. The perceptions of the respondents as to their strengths, weaknesses, successes and needs were highly useful in developing the topics for discussion at the Water Quality Management Expert Workshop. The cooperation of the utilities responding is greatly appreciated.

APPENDIX C: WORKSHOP ATTENDEES

WAER Q UALIMA AGEME HW S RUCURE I WIHI A WAER UILI			
articiants			**EMail**
Barsotti	Michael	Champlain Water District	mikeb@cwd-h2o.org
Brown	Miranda	Washington Suburban Sanitary Commission	mbrown1@wsscwater.com
Cummings	Ed	Zone 7 Water Agency	ecummings@zone7water.com
Douglas	Ian	City of Ottawa	ian.douglas@city.ottawa.on.ca
Gedney	Bill	Southern California. Water Co.	wcgedney@scwater.com
Hilmoe	David	Seattle Public Utilities	wcgedney@scwater.com
Knudsen	Mark	City of Portland Bureau of Water Works	mknudson@water.ci.portland.or.us
Lynch	Gary	Park Water Co.	gary@parkwater.com
Neden	Douglas	Greater Vancouver Regional District	dneden@grvd.bc.ca
Oblonsky	Sandy	SCVWD	sandoblo@scvwd.dst.ca.us
Parekh	Pankaj	Department of Water and Power	pankaj.parekh@ladwp.com
Quintanar	Daniel	Tucson Water	dquinta1.ci.tucson.az.us
Rudnickus	John	City of Toronto	jrudnick@city.toronto.on.ca.
Barrett	Slyvia	MWDSC	sbarrett@mwdh2o.com
roect Adv isory Committee			
Owen	Christine	Tampa Bay Water	cowen@tampabaywater.org
MacBride	Barry	Winnepeg	bmacbrid@city.winnipeg.mb.ca
Hunsinger	Ron	EBMUD	rhunsing@ebmud.com
Kawczynski	Elizabeth	AwwaRF	ekawczynski@awwarf.org
roect e am			
Means	Ed	MEC	emeans@mcguireinc.com
LeChevallier	Mark	AWWSC	mlecheva@amwater.com
Reich	Ken	MEC	kreich@mcguireinc.com
Frey	Michelle	MEC	mfrey@mcguireinc.com
McGuire	Mike	MEC	michaelj@mcguireinc.com
Ospina	Lorena	MEC	lospina@mcguireinc.com

APPENDIX D
WATER QUALITY BEST PRACTICES

The following represents a potential checklist of positive features or attributes of utilities that were identified in the Expert Workshop. They can serve as guidelines or "best water quality management practices" to encourage in water utilities. In contrast, Appendix E contains "signposts of dysfunctionality". If you observe Appendix E issues in your utility, they represent a potential for poor water quality performance and should be closely scrutinized.

- ❑ The utility has a documented water quality policy. The policy is board adopted and supported by the organization (water quality is considered a core business function). There is a water quality ethic in the organization where employees know their respective roles/accountabilities. Realistic water quality goals are included.
- ❑ Water quality is included as a specific element of the utility's overall strategic plan document.
- ❑ Customers are satisfied with their water quality. All customers get high quality water that surpasses standards. The utility actively communicates with the customers.
- ❑ There is full compliance with water quality regulations and requirements.
- ❑ The utility voluntarily seeks to minimize environmental impacts of utility/water quality activities.
- ❑ The utility is sensitive to special customer needs.
- ❑ The utility is responsive to water quality matters (internally and externally).
- ❑ The water quality function in the utility is recognized as a stakeholder in organizational decisions.
- ❑ Waterborne disease outbreaks are absent in the community.
- ❑ The utility recognizes and contributes to the public health good of the customers through water quality improvements.
- ❑ The utility is recognized by the regulatory agency as being progressive and responsive.
- ❑ There is a positive image and a demonstrated public/media confidence.
- ❑ The utility is recognized as a progressive utility (industry, media, colleagues, internal & external)
    - Share knowledge in industry
    - Recognized mentor
    - Support research
    - Encourage staff participation in conferences, technical exchanges, industry activities, etc.
    - Innovative
- ❑ The utility is able to attract and retain WQ staff.
- ❑ The community observes decreasing (leveling off) bottled water sales.
- ❑ The utility has documented strategies to meet or surpass future regulations (investigations, research, plans, etc.).
- ❑ The utility is engaged in legislative/regulatory process.
- ❑ The utility responds to a crisis in an organized, calm fashion. Its response plans are effective and timely.
- ❑ The utility uses cost-effective solutions to produce high quality water.
- ❑ The utility resists complacency.

- Proactive
  - Engaged in issues
  - Motivated
  - Comfortable operating their plants
  - Proud of their plants, etc.
  - Certified operators
  - Training is valued
  - Actively seek accreditation
  - Use of Partnership for Safe Water (or other programs to improve water quality/operations)
- There is documented water quality reliability
  - Multi-skilled staff
  - Robust treatment
  - Redundant processes
  - Emergency response
- There is open and easy access to data and information.

APPENDIX E
WATER QUALITY MANAGEMENT EFFECTIVENESS

Background

The Expert Workshop participants identified dozens of examples of water quality management effectiveness in the following categories:

- Regulatory compliance
- Water quality goal setting
- Strategic planning
- Communications
- Organizational development
- Key relationships/public involvement
- Monitoring and analysis

These examples provide potential opportunities for the progressive water utility to improve its management of water quality. It is clear there is no "one way" to manage this function but many of the following examples were common to some of the best run water quality functions in U.S. water utilities.

Regulatory Compliance

The management goal should be to establish a clear chain of responsibility for water quality regulatory compliance. To this end, the water utility should:

- Clearly define responsibility/accountability.
- Establish a "Compliance Officer" for internal audit and organization of compliance information (defined protocol to prepare and track compliance & reporting).
- Identify individuals responsible for permitting/plan review communications.
- Identify individuals responsible for operational/SDWA compliance reporting.
- Provide focused communications to State compliance authorities.
- Demonstrate this accountability on an organization chart; defined on a personal basis in meetings.
- Ensure that the reporting policy is understood (this may or may not include documentation as some private utilities avoid documentation from a legal liability standpoint).
- Designate a Regulatory Compliance Manager & develop a clear job description
- Have understood chain of command to escalate issues/reporting.
- Provide conservatism in operations (follow up on slightest detection, use early warning detects <MCL as warning sign).
- Prepare a written monitoring schedule (possibly hand held computer-enabled).
- Establish "compliance teams" (cross-functional) to meet specific regulations (e.g. TCR Action Plan, THM Action Plan, NDMA Action Plan). Develop a charter for each team. Prepare a written action plan (e.g. MTBE action plan that board adopts).
- Establish a water quality responsibility checklist for the organization (things that must be accomplished by the teams). Require signoff when each item is completed.

- Minimize points of contact with the regulator – focus on one person.
- Establish specific water quality goals.
- Define the reporting relationships (e.g. follow-up requirements for coliform speciation).
- Document sampling protocols/procedures (consider bar codes on sample taps to ensure sample integrity (i.e. that sample indeed came from that tap). Also records time of sample collection.
- Provide online monitoring with alarm.
- Provide early warning capabilities for source water and distribution system locations.
- Ensure there are dedicated resources to support the analyzers and shift the operational culture to one of trusting the online measures (transparency of calibration and maintenance activities).

Water Quality Goals

Water quality management should establish specific goals and routinely measure performance against those goals. The organization must participate in establishing the goals and support their implementation. Specifically:

- Define internal "Action Levels" for Response by Water Quality.
- Investigate any positive (TCR goal is 1%). Detailed follow-through is essential to attempt to determine the source of the problem. This could include sample error, inadequate treatment, flow changes, disinfectant residual problems, regrowth, etc.). Work to establish a culture of "inquisitiveness" and responsiveness in reacting to potential indicators of pending water quality problems.
- Water quality must be a shared value across the organization (especially within operations).
- Consider developing goals through a Water Quality Committee, which includes all organizational functions.
- Goals are monitored and reported out every month, when trend line goes up every month the issue is escalated to process engineering group to investigate.
- Develop a water quality program document including mission and objectives.
- Establish goals with financial disincentives for not meeting goals (especially in contract ops). Develop general long-term goals and numeric short-term goals.
- Produce graphical format trend to be more easily understood by staff.
- Include phased "trip wires" - Consider multiple escalating targets that trigger different actions. (e.g. particle counts as warning for turbidity >3 um, >10 um, <0.1 NTU all filters).
- Seek consistency of operations
    a. 100% of the time <0.20 NTU
    b. 95% of the time <0.15 NTU
- Try to keep the goals simple and understandable.
- Participation in the Partnership for Safe Water.
- Be sensitive to how goals are perceived within the organization, they are not regulations.
- Employee performance should be measured based on actions towards meeting the goals as opposed to meeting the goals specifically.

Planning

Conduct formal water quality strategic planning. The organization must share the vision. The planning should:

- Create a vision.
- Establish goals.
- Engage the other organizational stakeholders to help ensure organizational buy-in.
- Include development of a capital improvement plan for water quality improvements in the source water, water treatment plant(s), distribution system, monitoring, and research areas.
- Include creation of a short-term plan including budget, staff, resources, sign-posts/milestones/action items/metrics/measurables, performance measures (staff accountability).
- Consider strategies to gain public acceptance for water quality initiatives outside of regulations.
- Consider pending and future regulations.
- Integrate water quality into the organization wide strategic planning .

Communications

Transparent and timely communications within and external to the organization is essential to a managing the water quality function in a water utility. To this end, the utility should:

- Provide routine internal and inter-disciplinary meetings to exchange information, solve problems, build awareness, prevent miscommunication and build staff relationships. These meetings can enable Senior Management to "Take the Pulse" of the water quality group.
- Consider an annual water quality team retreat to step back and regroup.
- Provide routine reports that summarize activities and planned events at defined frequencies (e.g., daily, weekly, monthly).
- Acknowledge differences in personnel and their communication style and preferences.
- Use conference calls to engage field personnel.
- Use a structured review or update of all components so that all staff has to contribute something to the meeting.
- Establish email rules (e.g. information only, not decisions, and keep it short, no engineering by email; problem solving or conflicts resolved face-to-face, not in email).
- Provide exceptions reports to highlight any water quality event outside of defined norms.
- Incident analysis – used to understand, diagnose and resolve operational or water quality abnormalities. Requires using data to learn about system behavior, and ultimately affects Standard Operating Procedures, reaction to trends, etc.
- Consider a "buddy system" –find an analogue person in another utility to help with trouble-shooting.
- Consider "Mutual Aid" agreements with neighboring utilities.
- Set up a Board Committee with a water quality focus – educate the committee members and they can act as advocates for water quality on the Board.
- Establish teams: chemists, microbiologists, engineers, limnologists. Build internal advocacy capabilities and promote staff to advocate to their management.

- Provide daily system status reports for the organization and key customers (wholesaler selling to retailer).
- Involve others in writing conference papers; provide a more general presentation for the board.
- Consider thermal arrays in reservoirs to capture temperature profiles (thermistor array).
- Provide for an "open door" policy with the board. Entertain any water quality issue.
- Create a "home" for water quality issues. Have a specific committee that has responsibility for water quality policy matters.
- Have the governing Board meet at the plant/lab to show it off.
- Produce and distribute a twice per year water quality report to the board. Issues 'in-between" can come in special reports. Explain how things are handled, steps taken, how we do water quality business.
- Have Board adopt water quality goals. This will send a powerful message to organization about importance of goals.
- Take issues to the board early.
- Conduct weekly executive staff meeting reviews water quality (among other things)
- Conduct timely communication to management on developing water quality issues or problems.
- Support the General Manager at the national level (where applicable)
- Establish goals; create vision; develop consensus.
- Create multiple mechanisms for communication- databases, email/intranet/website, journals, committees, activity reports, presentations/in-services, brown bag seminars, focus groups, surveys, etc.
- Define/understand your audience-regulators, customers, other departments, governing board, employees, press/media.
- Establish/Create/Nurture/Build a 2-way dialogue.
- Develop a means of evaluating the effectiveness of your internal and external communications.
- Monitor treatment plant data (in one example a "water quality committee" follows the information).
- Conduct weekly operations meetings.
- Conduct telephone conference call (snowflake to tap) – everyone attends when there is a problem. People know that if there is a problem they need to be on the call and prepared to share information.
- Provide liaison with groundwater basin groups and other users/stakeholders of water to share information re: water quality. Trend to watershed wide data sharing.

Organizational Development

The development and training of staff is vital to sustaining a water quality ethic and culture in a water utility. Moreover, the rapid technological changes in the analytical chemistry and microbiological areas (as well as process engineering) require staff to stay current. Utilities should:

- Conduct training: Technical and communication (knowledge transfer). Emphasize sampling training and refresher programs.
- Provide career development opportunities.

- Encourage workforce flexibility- provide cross training in the laboratory where practical and feasible.
- Provide financial incentives to get higher grade levels (beyond the minimum required).
- Ensure succession planning. Actively identify and cultivate future water quality management leaders as well as technical staff.
- Encourage continuing education, workshop and conference attendance.
- Empower employees to make decisions; trust them.
- Establish a mentoring program.
- Recognize the need for competitive recruitment; adopt a progressive attitude towards seeking and retaining talent. This includes providing competitive salaries.
- Provide performance evaluations and feedback mechanisms.
- Align individual staff plans with the water quality goals of the organization.
- Integrate water quality Staff into Other Utility Business Areas.
- Create team liaisons to bring the WQ perspective.
- Formalize integration by staff resource planning upon project initiation and discipline "sign off" at the end of the large projects.
- Have water quality engineers rotate at plants to help solve problems & work with operational staff.
- Assign a water quality engineer to shift work three weeks per year to build relations and knowledge between engineers and operators.
- Continuously train staff on water quality issues such as research activities, regulations, SOPs Build more multi-disciplinary skills within water quality.
- Train operations staff.
- Participate in the Partnership for Safe Water.
- Conduct applied research activities to build in-house knowledge and credibility, especially in moving to implementation. This builds awareness, communication and forces internal problem solving. It also serves as a model for improving other operational aspects of the system.
- Establish a method to ascertain factual basis of problem event to allow learning within the organization.

KEY RELATIONSHIPS/PUBLIC INVOLVEMENT

In cultivating key relations and providing a uniform/consistent public interface:

- Develop professional relationships with regulatory authority by having structured, routine meetings to exchange information.
- Routinely meet with the local health department to build an alliance on water quality initiatives.
- Position the water quality group to be perceived positively by the utility as a technical resource and a group that is responsive to other groups' needs.
- Provide for a substantial interface between water quality and the customers/media. Choose your spokesperson carefully and train them in media interactions. Having local health department endorse utility actions is a real plus.
- Establish a water issues training forum for the general public.

MONITORING & ANALYSIS

- Find cost-effective use of lab services – including balancing internal/ external services with monitoring program design.
- Embed ownership of the process lab function in operations. Water quality should provide a quality assurance function, including training.
- Have the lab/water quality "own" the sample collection function for quality data.
- Require WQ staff (either WQ inspectors, operators or lab personnel) to handle customer inquiries and follow-up.
- Provide for an off-hour, on-call capability for this function.
- "Pre-qualify" acceptable contract laboratories rather than have purchasing control the contracting.
- Provide customer information.
- Consider a Web-based system with auto monitors.
- Consider on-property water quality (monitoring on buildings after the meter), allows consumers to make decisions about POU/POE- do side-by-side comparisons of tap and bottled water.
- Look at consumer inquiries as an opportunity to make them feel better – collect samples to calm them.
- Conduct threshold odor samples every shift from reservoir settled water and bottom of filter. This ensures some level of understanding regarding when problems were developing to trigger investigation. Use whatever taste and odor method works best.
- Observe reservoir bloom conditions as early warning signs of potential problems. triggers field Flavor Profile Analysis.
- Involve your team. There is often a tendency to only let the "highest level of smart people" review the results. Most successful monitoring program engaged the sampler, the lab analyst, the water quality engineer and the water quality manager (as well as operations folks). Labs used to be involved only with generating and interpreting the result, movement away from this now.
- Have someone dedicated to making sure the on-line data/information is accurate.
- Assign an information tech to the water quality group. Ownership of online information can be a problem.
- Provide comparable WQ throughout distribution system, while maintaining level of service.
- Institute a program to "exercise" storage in the reservoirs (treatment/transmission specialists). Specialization has "de-linked" treatment and transmission.
- Involve the operators. Many treatment operators feel isolated from the distribution system.
- Reduce barriers to teamwork.
- Establish SOPs for notification of special customers re: quality changes.

APPENDIX F
CHECKLIST FOR PROBLEM ORGANIZATIONS

Organizations that are not structured properly for optimal management of the water quality function invariably show visible signs of this dysfunctionality.

As a manager of this function you should look at the following issues as signs of dysfunctionality. Some are relatively benign but others are indicative of an organization that has seriously compromised its ability to manage water quality in the best interests of the customer or the utility.

These signs include:

- ❑ The water quality function has become so specialized that they do not recognize the operational importance of the data they are generating. They generate data, rather than interpret data andgenerate information.
- ❑ Employees are afraid to report water quality violations or problems
- ❑ Employees are reporting violations to external entities before management knows about it
- ❑ Employees don't understand what is a problem or when to report it.
- ❑ There are no specific goals/objectives, or standard operating procedures to ensure water quality.
- ❑ Communications frequently breakdown within the water quality function or between functional units of the organization.
- ❑ No follow up (policy changes) after incidents. Incidents are not used as a learning opportunity.
- ❑ Operating inconsistencies occur (night to day shift differences).
- ❑ There is unclear delegation of authority for water quality or treatment process decisions.
- ❑ There is a distrust of water quality information.
- ❑ There is denial of information (it can't be right).
- ❑ Everything is an emergency (response doesn't match event).
- ❑ Walls (territories) built between functions.
- ❑ Systems are brought on line with inadequate training and testing.

- Systems are brought on line on Friday.
- There is "throwing it over the cubicle wall".
- Someone that doesn't ask for help when they need it.
- There is dissention and distrust among employees.
- There are large number of employee grievances.
- The prevailing attitude is "That's the way we've always done it."
- There are frequent plant upsets (turbidity spikes etc.).
- There is a large amount of emergency call-outs & overtime.
- There is inadequate WQ involvement in bringing new facilities on line.
- Water quality functions are chronically under funded.
- Water quality is not viewed as a "team player".
- Water quality is not team player.
- There is tension between Engineering, Operations and Water Quality (attain functionality, operability, reliability, maintainability, Safety (FORMS).
- Online instrumentation is not maintained and is viewed as unreliable.
- The operators feel there are excessive SCADA alarms.
- There is a high accident rate.
- Upper management doesn't understand budget needs (every year is a battle). Water quality is singled out consistently.
- Employee performance reviews are not done or are inconsistently performed.
- All talk no action; too much time wasted in meetings.
- Blame storming is common.
- There is constant revisiting of decisions.
- Decisions aren't timely or not made at all (analysis paralysis).

- There is inadequate evaluation of solutions prior to implementation.
- There is no log book (lack of documentation).
- There is high personnel turnover.
- Production and distribution operating in silos.
- Intentional falsification of records.
- Tendency to do the minimum.
- Newspapers have comics about the quality of the water.
- Regulator, Board or GM hears about your problem from the media.
- Regular QA/QC busts.
- Information is not timely.
- There are frequent lapses in security (reservoirs, etc.).
- Employees following the letter of the law and not the spirit of the law.
- Regulatory surprises occur.
- Not being involved in the utility's water quality, organization and master strategic planning.
- The organization frequently reorganizations.
- Water quality decisions are not communicated/supported by consecutive systems.

APPENDIX G
AMERICAN WATER SYSTEM APPROACH

Valuing Research – American Water System Approach

A survey was conducted of American Water System Water Quality Directors and Managers of Operations to assess their perceived benefits of research. Additionally, an analysis was conducted of selected research projects where the cost benefit could be determined based on subsidiary documented cost-savings.

Summarizing all year-to-date direct (labor and overhead) and expenses (supplies) costs results in the annual research cost billed to American subsidiaries. For the year, 2000, expenses charged to American subsidiaries for research authorizations totaled $1,404,295. As demonstrated in Figure G.1, nearly all these expenses were for labor and overhead (overhead at the Belleville Laboratory is approximately 170%). Not included in the total is $463,529 that was credited from various research reimbursements. These reimbursements offset all but $76,993 in expenses, reducing this amount to only 4% of the overall research cost. Therefore, the *total cost* of research in 2000 was $1,867,824, but American subsidiaries only paid for $1,404,295, a 25% reduction in expenses.

For the accounting summaries, it is not possible to precisely detail the expenses for the Finished Water *Cryptosporidium* Monitoring project because these charges were mixed with all other Belleville Lab water quality activities. It is *estimated* that expenses were in the range of $500,000 for labor (and overhead) and $200,000 for expenses for this project.

The benefits from externally funded research are much greater:

- Competitive research funding ensures that American's research program is first-rate. There's no reason for American subsidiaries to support second-rate research.
- External research funding has allowed some of the world experts to collaborate with American researchers to bring their skills to bear on research issues within the American System. There is no cost to American subsidiaries for experts' time, research, or analysis (these are paid by the external funding agency).
- Collaborations have allowed American researchers to acquire new skills to meet American subsidiaries informational needs (examples: integrated membrane research, *Cryptosporidium* cell culture, particle counting, *Mycobacterium* analyses, etc.). These skills can be applied in future projects meeting subsidiary needs.
- The collaborative research provides additional information beyond that developed by American researchers.

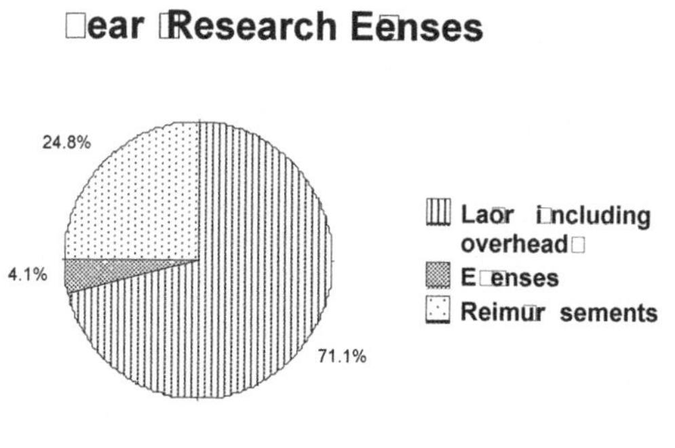

Figure G.1 Summary of research expenses, including reimbursements.

The impact of this last point can be quantified in terms of *research value* compared to *research cost*. For the 40 externally funded research projects conducted since 1987, American received $16.5 million of research information for $4.7 million in in-kind cost. Therefore, American rate-paying customers received $3.51 in water quality information for every dollar spent on these externally funded research projects.

Because of the reputation of American's research, many scientists seek out American for participation in their research studies. Additionally, American scientists will contact colleagues to offer our participation when the work may be beneficial to American subsidiaries. Examples include on-going studies of endocrine disruptors, for which 50 sites will have raw and treated water examined for estrogenic agents. This study will permit a determination of the impact of this emerging health issue on American operations and allow subsidiaries to authoritatively communicate these issues to customers – with virtually no cost. Other projects include application of new taste and odor assays, analysis of algal toxins, *Legionella* testing, and treatment plant modeling. If these preliminary studies yield useful results, further investigation is possible through collaborative research projects.

It is important to note that many Water Quality projects cannot be quantified in terms of a cost/benefit. Many projects are focused at meeting the informational needs of the operating companies, improving water quality, or meeting new drinking water regulations. For example, how has American's *Cryptosporidium* research benefited American subsidiaries to reduce treatment costs? It probably has resulted in increased costs (through higher coagulant doses and tightened plant operations), but uniformly the *Cryptosporidium* research is cited as an example where American's research has excelled (see the April 2001 survey results). Likewise, few utilities would have identified *Cryptosporidium* as an important research topic in 1987 despite the outbreak it caused in Carrollton, Georgia (the chlorine resistance of the oocysts and the conventional nature of the Carrollton treatment process made it clear to System Researchers the importance of this research topic).

In the mid-1980's, a number of American facilities were having problems with frequent coliform violations. Research has identified the causes of coliform regrowth and outlined a number of approaches to combat the problem. Today, coliform violations are rare and typically result from unusual circumstances. What is the value of coliform compliance? Undoubtedly there is a monetary value in terms of customer satisfaction, regulatory relief, and new business opportunities; but exact quantification of this value is difficult.

There have been, however, a number of research projects that have been directly aimed at lowering drinking water cost, or improving efficiency so that cost increases can be minimized. The results of the analysis are summarized in Table G.1.

Table G.1: Summary of Cost Benefits of Specific Research Projects

Company*	Annual Savings	Research Cost	Return on Investment (months)
Energy Management			
PA-Am, Hays Mine	50,000 - 60,000	41,000	8 - 10
KY-Am	60,000	7,800	2
MO-Am	10,000 - 15,000	4,000	3 - 5
Meter Reading			
IL-Am, Interurban	148,000	48,000	4
Corrosion Control			
IN-Am	138,161	16,985	2
VA-Am, Hopewell	10,653 - 17,901	8,435	6 – 10
TN-Am	21,000	12,000	7

* Benefits are cited for six other systems not listed in this table.

The average time for the subsidiary to recoup their research investment was 5-6 months. Rather than prove that research lowers the operational cost for all systems, the data in Table G.1 illustrates that cost savings are possible and suggest that if these savings are possible for some systems, they probably are likely for most American subsidiaries. Even 0if the data in Table 1 overestimate the research benefits by a factor of 10, a 60-month return on the research investment would still be equitable. Examined another way, nearly a quarter of the annual cost of American's research can be justified based on the cost savings reported in Table 1 alone.

Many questionnaire respondents cited the positive benefits from the high visibility and good reputation that American's research provides to the company. This reputation has facilitated interactions with State and Federal agencies and improved customer relations. Research data on particle counting, ground water monitoring, *Giardia* and *Cryptosporidium* analyses, etc. have been instrumental in providing State approval for various subsidiary projects. In one recent example, the Illinois EPA permitted Illinois-American latitude in dealing with treatment plant residuals because of American's research on wash water recycle. American Water Services reports using System Research expertise as a marketing tool for new service contracts. The development of a patent for *Cryptosporidium* detection resulted in a license agreement between American Water Resources and Dupont-Qualicon.

The scope of American's research is based on the development of the 5-year Research Agenda. The agenda is developed from a survey of System Company Presidents, Operation Mangers, facility Managers, and the Directors of Engineering and Water Quality. This information is provided to a research advisory group comprised of representatives of Engineering, Water Quality, and subsidiary Management. The Research Agenda is updated every three years and an analysis provided in the last Research Agenda (1999-2004) showed that 80% of the projects identified in the previous agenda had been at least partially addressed or were in progress. Changes in prioritization, lack of resources, or development of new priorities accounted for the remaining 20% of the projects. However, the nature of American's business is rapidly changing and the development of the next update, scheduled to begin later this year, will provide an opportunity for subsidiaries to refocus the research program.

The water quality management function will continue to conduct surveys of American subsidiaries to ensure that the technical services and intellectual resources continue to add value, meet needs, and provide comprehensive, timely and cost-effective solutions.

Contracting for Services

Water quality management can also benefit from consolidation – either through mergers and acquisitions - or through contractual arrangements. One of the greatest benefits for American Water is the network of water quality professionals available for consultation. The image of the water utility as a "stand alone fortress" totally self contained and self-providing is long gone. Today it's the network of services, from out-sourcing water quality functions to networks of service providers.

American Water has consolidated its customer service (into a national call center), accounting, and IT services. Within the water quality function, they have water quality specialists that are circuit riders between a number of plants in some areas. This is common in Pennsylvania where one person may cover 5 systems within a geographical region. These small communities are able to share the costs of having a highly qualified individual manage the water quality function - and this person is plugged into a state-wide (and nation-wide) network.

Contract services can be an area where more systems benefit from obtaining needed water quality professionals. This could range from a formal contract service, to a close working relationship with a consultant to provide that "high end" overview and control.

REFERENCES

Aspan, Howard N., Summer 2000, *Running in Nonconcentric Circles: Why Environmental Management Isn't Being Integrated into Business Management*. Environmental Quality Management.

Auluck, Randhar, 2002, *Benchmarking: A Tool for Facilitating Organizational Learning*. Public Administration and Development

Australian Guidelines for Water Quality Monitoring and Reporting: Summary 2000. National Water Quality Management Strategy; No. 7A. Australian and New Zealand Environment and Conservation Council and the Agriculture and Resource Management Council of Australia and New Zealand.

Cromwell, J., Owen, D., Dyksen, J., and E. Means. 1997, *Managerial Assessment of Water Quality and System Reliability*, AWWARF.

De Bakker, Frank et al, 2002, *Responsible Chain Management: A Capability Assessment Framework*. Business Strategy and the Environment.

Frooukh, Luay M., 2001, *Decision-Support System for Domestic Water Demand Forecasting and Management*. Water Resource Management.

Garvin, David A., 1993. Building a Learning Organization. Harvard Business Review, July-August 1993, pp. 78-91

Gilmore, Audrey et al., 2000, *Managing Strategic Change in Small and Medium-sized Enterprises: How do Owner-Managers Hand Over their Networks*. Strategic Change.

IRS, 1998, *IRS Management Review: Using Human Resources to Achieve Strategic Objectives*. Learning Strategies Review Issue 8. IRS Eclipse Group: London.

International Organization Standards: ISO 9000. Quality Management Principles. 2002.[Online]. Available: <http://www.iso.ch/iso/en/iso9000-14000/iso9000/qmp.html <http://www.iso.ch/iso/en/iso9000-14000/iso9000/qmp.html> > cited June 1, 2004.

Naumann and Associates. 2002. "What is Customer Value ?"[Online] Available: <http://www.naumann.com/ii.htm <http://www.naumann.com/ii.htm> > cited June 1, 2004.

NHMRC/ARMCANZ Coordinating Group, 2001, *Framework for Management of Drinking Water Quality: A Preventive Strategy for Catchment to Consumer*, Public Consultation Document,

Schein, E. 1997, *Organizational Culture and Leadership*. Jossey-Bass: San Francisco, CA.

Scott, F.G., 2001, *Beyond Leadership –Managing the Business of Public Works.* AWWA Trends in Water Series- Excellence in Action: Water Utility Management in the 21st Century.

Smith, KD, Taylor, WGK., 2000 *The Learning Organization Ideal in Civil Service Organizations: deriving a Measure.* The Learning Organization, 7 (4):194-205.

ABBREVIATIONS

AwwaRF	Awwa Research Foundation
CDC	Center for Disease Control
EPA	Environmental Protection Agency
ISO	International Organization for Standardization
IT	Information Technology
LIMS	Labortory information management system
MCL	Maximum contaminant level
MTBE	Methyl tertiary-butyl ether
NDMA	N-nitrosodimethylamine
NTU	Nephelometric turbidity units
PAC	Project advisory committee
POU/POE	Point of use/point of entry
SDWA	Safe Drinking Water Act
SWTR	Surface Water Treatment Rule
TCR	Total Coliform Rule
THM	Trihalomethanes
USEPA	United States Environmental Protection Agency

IWA Publishing's authorised EU representative for General Product Safety Regulations is Diane D'Arras, 15 rue Duret, 75116 Paris, France, e-mail: safety@iwap.co.uk.

Printed and bound by CPI Group (UK) Ltd, Croydon, CR0 4YY
17/09/2026
02218804-0011